IN THE BLOOD

IN THE BLOOD

BLOOD

UNDERSTANDING AMERICA'S
FARM FAMILIES

ROBERT WUTHNOW

PRINCETON UNIVERSITY PRESS
PRINCETON AND OXFORD

Copyright © 2015 by Princeton University Press

Published by Princeton University Press,
41 William Street, Princeton, New Jersey 08540
In the United Kingdom: Princeton University Press,
6 Oxford Street, Woodstock, Oxfordshire OX20 1TR

press.princeton.edu

Jacket/Cover photograph: Wind Turbines, © Stacey Evans

All Rights Reserved

First paperback printing, 2020
Paperback ISBN 978-0-691-21072-8
Cloth ISBN 978-0-691-16709-1

British Library Cataloging-in-Publication Data is available

This book has been composed in Adobe Garamond Pro, Ristretto Slab Pro, and Ultra

CONTENTS

INTRODUCTION – 1

1 – FAMILIES – 12

2 – NEIGHBORS – 46

3 – FAITH – 72

4 – INDEPENDENCE – 95

5 – THE LAND – 119

6 – TECHNOLOGY – 140

7 – MARKETS – 163

AFTERWORD – 185

APPENDIX – 191

NOTES – 199

INDEX – 219

IN THE BLOOD

INTRODUCTION

Here in corn and soybean country the land stretches endlessly to meet the sky in all directions. Vast acreages of grain spread across gentle rises and shallow valleys. A row of tall electrical poles leads off the highway down a sanded country road toward a farmstead surrounded by trees. Corn ripens on one side and cattle graze on the other. Pungent goldenrod lining the fencerows scents the warm late summer air.

A left turn into the driveway reveals a modest two-story, hip-roof house cased in white aluminum siding. A thick evergreen shelterbelt protects the house on the north. Thinly spaced elms to the south permit ample sunlight during the winter. Near the back of the house a path to the swing set looks well used.

Farther along the driveway a double garage stands near a faded red barn and behind it a large metal machine shed. The machine shed is noticeably newer than the barn. Toward the end of the driveway a giant self-propelled spraying rig with upright folding booms flanks an outlying clutter of half-rusted implements from earlier days.

Neil and Arlene Jorgensen have been farming in this part of the country all their lives. They are fifth-generation farmers. Mr. Jorgensen's parents, Clay and Mary, live a mile to the east and a quarter mile south. Clay's great-grandfather purchased the family's first quarter section here in the early 1880s. Clay's grandfather built the house and barn where Clay and Mary live. "I'm still sleeping in the same bedroom I was born in," Clay says.

Families like the Jorgensens are the backbone of America's rural economy. Many of these families have farmed in the same location for generations. In some areas they coax the nation's corn and soybeans toward harvest, in other places they nurture its wheat, and in still others they tend its cotton. Their daily labor supplies the milk we drink and the fruits and vegetables we eat.

Family relations are integral to the Jorgensens' farming activities. Neil and Clay farm in partnership. The two generations draw income from the same crops. Although Clay is old enough to have retired, he stays active running errands, helping feed the cows, manning one of the tractors during planting, and driving the truck during harvest. Neil does the heavy field-work and handles most of the management decisions. Hardly a day passes that the two do not spend time working together.

The Jorgensen women are as actively involved in farming as their husbands. Arlene has a job in town but does most of the farm bookkeeping. She and Neil decide together on major purchases, such as land and equipment. Mary looks after the grandchildren. "I'm the go-getter," she says, explaining that she runs errands, drives a tractor, and brings meals to the field during harvest. Neil eats at his parents' house about as often as at his own.

Both couples are proud to be doing what their ancestors did. They consider it fortunate to be living near each other and working together. The physical labor is not as exhausting as it used to be. The tractors are bigger and better. Information technology has dramatically changed the way farming is done. The Jorgensens no longer raise hogs. Corn and soybean prices have been good the last few years.

The Jorgensens are also facing challenges. When Neil was growing up, it seemed natural that he would farm. He started helping with the chores in grade school and was driving the tractor by the time he was in junior high. "I guess I've got farming in my blood," he says. He hopes one of the children will follow in his footsteps but is unsure if that will happen. It has been harder to pass his knowledge on to his sons and daughters and to save enough to get them started. Machinery is almost prohibitively expensive. The new combine he purchased three years ago cost a quarter of a million dollars.

Relationships with the neighbors have been changing too. Clay remembers when neighbors shared machinery and got together to visit on Sunday afternoons. Now that he is almost retired, he meets a couple of other farmers his age for coffee early on weekday mornings. Neil is too busy. Besides that, there are hardly any farmers nearby. Only the ones with large tracts of land are left. Neil worries about being squeezed out before he is old enough to retire. The competition is fierce.

Then there are the challenges of keeping up with new technology. The high cost of machinery necessitates careful budgeting. Seed is now genetically engineered and costs ten times what it did a decade ago. New fungicides and pesticides come with confusing instructions. Too much at the wrong time will stunt the grain. Information technology makes it easier to stay current of new developments but also makes it important to keep up with market fluctuations.

A century ago approximately six million Americans farmed. That number has declined dramatically. According to the US Census Bureau fewer than 750,000 employed Americans list their principal occupation as farmer, meaning that they earn their primary living from farming land they own or rent. If people who describe themselves as farm managers are included, the total rises to about a million. More Americans earn their livings as accountants than as farmers. Twice as many.[1]

Although farmers are a small fraction of the US labor force, farming continues to be a topic of interest and importance to the nonfarming public. One reason is that nearly everyone interacts indirectly with farming three times a day—at breakfast, lunch, and dinner. The food supply depends on farming. We expect food to be there when we want it, and we expect it to be healthy and reasonably priced.

A second reason is that American history is rooted in farming. It is hard to understand America's past without considering the central role of farming to leaders such as Thomas Jefferson and Andrew Jackson and among the millions of pioneers who settled the land. Many Americans who now live in cities and suburbs hale from farm families.

The cultural legacy of farming generates continuing interest in understanding the experiences of people who live close to the land. A person interested in literature does not have to look far to find accounts by writers who have left the city, returned to the family farm or purchased a small parcel, and described their experiences raising animals and rediscovering the serenity and challenges of rural life.

A third source of interest stems from the fact that rural America is vitally important to the nation's public policy. What farmers do with the land they farm has important implications for environmental and energy policies. How agriculture is affected is an important consideration in international trade negotiations. It is a frequently contested issue in policy discussions about food stamps, school lunch programs, and public health.

Farming is also of interest and importance in academic discussions about the nature of society. Theories of society that emerged toward the end of the nineteenth century emphasized the large-scale shift from agrarian-based to industrial-based economies. With farm life declining, long-held traditions and values were assumed to be diminishing as well. Scholars expected the close-knit relationships that characterized farming communities to be replaced by something better suited to urban life. Gender relationships would probably change. Even religious beliefs and practices might change.[2]

Questions about social change have generated continuing interest in the differences between rural and urban life. Much of the attention has focused on the growth of cities and suburbs. The related questions have to do with changes in rural areas. These questions concern the impact on farm life of such changes as declining population in farming communities, the aging of farm families, succession of farms to the coming generation, and the effects of changes in technology and markets.

Perhaps because it is of such widespread interest, farming is a topic that sometimes eludes clear understanding. Stereotypes of farming range from depictions of country bumpkins living old-fashioned lives to images of rural plutocrats reaping undeserved benefits from the government dole. Stereotyping of this kind places farming outside the mainstream of modern middle-class America. Other stereotypes put farming too centrally inside the American story, attributing virtues and values to farmers that are somehow harder to find in urban locations.

Reliable information about farming comes from several sources. The news media carry stories about farm accidents, how the weather is affecting crop yields and food prices, and what the latest farm bills include in terms of government subsidies and regulations. The US Department of Agriculture (USDA) provides a wealth of information about crops, yields, prices, and the economics of farming. Agricultural economists, rural sociologists, and anthropologists have examined variations and changes in farm practices.[3] Fictional accounts and literary essays offer imaginative interpretations of rural life. Historical studies chronicle how farm families lived and worked in the past.[4]

The missing piece is what farmers themselves have to say about their lives. Why is farming important to them? What do they mean when they say farming is in their blood? How does the business of running a farm affect their families? Their relationships with neighbors? Their religious faith? Their sense of who they are as persons? Their understanding of the land? How are all of these understandings changing as farming changes?

THE PRINCETON STUDY

The research presented here was conceived of as a way of letting farmers themselves speak about their lives, telling their stories, describing their day-to-day activities, and talking about their families and their communities and the challenges they face as well as the opportunities they envision for the future. The idea was to prompt conversations by asking questions about various topics and then allow the conversations to develop in their own ways.

The study aimed to capture the voices of farmers who are seldom heard in any forum outside of farming communities themselves. Farmers who spend their days planting soybeans or wheat or harvesting corn or cotton or feeding livestock and milking cows. Farmers who may be earning a good living and farmers who worry about meeting the payments on their loans. Ordinary farmers like the Jorgensens whose stories would be missed in news headlines and government statistics.

The research was designed to record the stories of people who actually live on farms and who earn their primary income from farming. The researchers who collaborated with me on the project and I did not include people who may have lived in rural areas but who did not farm or people who could be described as hobby farmers because they earned their principal income from

4

working at some other job or from investments. We excluded corporations that owned or operated corporate farms but included farmers who may have formed family corporations or partnerships for legal and tax-related reasons. We focused on farmers who were engaged in what they considered to be family farming, whether that meant husbands and wives, siblings farming in partnership, or multigenerational farms.

The research design provided opportunities for farmers in several regions of the country and engaged in several different kinds of farming to tell us about their lives and to talk about the meanings and values they associate with their experiences in family farming. We talked with farmers like the Jorgensens who grow corn and soybeans and with farmers in other areas who specialize in wheat or cotton, who raise cattle, who operate dairies, and who specialize in fruits and vegetables.

In our interviews we asked farmers and farm couples to tell us about their daily lives and what they liked or did not like about farming. We asked how long their families had been farming in the area. They told us stories about their parents and grandparents. They recalled what it was like growing up on farms, if they had, and what adjustments they made, if they had not.

We spoke with farmers in their living rooms, at kitchen tables, in farm shops, and while they inspected crops and livestock. Some of the interviews were conducted with farmers by cell phone while they drove their tractors or hauled grain to town. Many took place on rainy days and during the winter months when work was slow. We talked with farm couples together and with farmers individually. Although the majority of our interviews were with men, approximately a third were with women and farm couples.

Farmers talked about the tough decisions they had made and how farming led to family conflicts as well as to family harmony. They discussed their neighbors and expressed their views about government policies. We asked that they speak candidly and say whatever they wanted to. We promised not to disclose their names or the names of their communities or to include information that might reveal their identity. Jorgensen is a pseudonym. Some of the farmers we spoke with lived in hip-roof houses, and some had swing sets. Their name was not Jorgensen.

The farmers we spoke with ranged in age from late twenties to late eighties. Most were in their fifties and early sixties, and nearly all were married. We talked with farmers whose families had been farming for three, four, and five generations. We also talked with farmers who had not been raised on farms or who were farming land that had not been farmed by previous generations.

In all, we conducted lengthy qualitative interviews with 250 people. Fifty were community leaders who told us their impressions of farm life from working closely with farmers as agricultural extension agents, as heads of local farm companies, and as clergy. The rest were farmers we contacted through a sampling design that ensured representation among small, medium, and large farms and in regions specializing in corn and soybeans, wheat, cotton, dairy, and truck farming. On average, the interviews took about ninety minutes. Many lasted two and a half to three hours. (The appendix provides additional information about the research.)

The comments and the stories told and the opinions expressed provide a rare opportunity to see how farmers view their worlds and to understand what farming means and why farmers consider it important. Information like this is not amenable to statistical generalizations. It requires paying close attention to the words and the speakers and their stories. The farmers we spoke with were not speaking as representatives of the farming population. They were describing their own experiences. It is from these descriptions and in the texture of the language itself that an understanding of their experiences can be attained.

Nearly every farmer we spoke with thought the public was misinformed about farming. Some blamed the media for telling stories that misrepresented facts about farm subsidies or that focused too much on bumper crops one year and crop failures the next. Some merely recognized that the nonfarming public purchases its food washed, processed, and conveniently packaged with only a dim understanding of how it originated on someone's farm. Many of the farmers we spoke with acknowledged their own responsibility for popular misunderstandings. It would be wonderful, they said, if people from the city could spend a day on someone's farm or if farmers could give talks to the public about farming. There was not enough time in the day for that to happen.

When we probed this concern about being misunderstood, we learned that farmers were not intent on communicating any one particular story that was not being told. They were not saying that the public had an overly glowing or romanticized view of farming and needed to be informed that farm life these days was a desperate struggle. Nor did they feel that farm life was a whole lot better than the public generally imagined.

Instead, the message that came through again and again was that farm life is complicated. It is more complicated than headlines or summaries from statistical surveys generally acknowledge. Its meanings and how farmers think about it vary not only from day to day but vary also depending on how a person looks at it. The good and the bad—the enjoyable parts and the ones that keep farmers awake at night worrying—are all woven together. As one of the farmers we spoke with put it, "There's always another side to the story."

Letting the different sides of the story come out—and indeed honoring the inevitable ambivalence present in the daily lives that any of us lead—is more difficult than it should be. It is easier to look for the simple headline or ask that the complexity be reduced to an argument that can be summarized in a single sentence. That kind of information is easiest to process even though we know from our personal experiences that nothing is quite that simple.

The cultural complexities of contemporary farming extend beyond the economic considerations that generally receive the most attention in food and farm policy discussions. The ambiguities or tensions involved reflect both the distinctive history of farming and its changing social location. Consider the following:

- Farming is a solitary occupation requiring long hours working alone and necessitating decisions for which the farmer takes sole responsibility, but farming is thoroughly embedded in social relationships that influence farming and change as farming undergoes change.

- Farming communities are tight-knit neighborhoods in which farmers share work and enjoy one another's company, but farmers' neighbors in these communities are uniquely their competitors in ways that characterize few other neighborhoods.

- Farming exemplifies the kind of traditional labor market in which decisions are made on the basis of ascriptive familial relationships rather than instrumental calculations, but farming has adapted to modern economic conditions in ways suggesting that rational decision-making processes prevail.

- Farming is an occupation that in many ways has changed very little and embraces values that emphasize tradition and continuity, but farming has also managed to adapt dramatically to new technologies that increase productivity and at the same time fundamentally change the social relationships in farm families.

- Farmers have a distinctly integral relationship with the land because of working closely with it on a daily basis, but this relationship is changing and perhaps becoming more distant as farmers employ larger equipment and use technologically advanced methods of farming.

- Farmers are thought to be particularly oriented toward religious values because of their dependence on the uncontrolled forces of nature, but questions must be asked as to whether this view is still correct as farmers have become more influenced by science, technology, and higher education.

- As the sole proprietors of small business operations, farmers are in a weak position with respect to global markets, and they realize this weakness and find ways to make sense of it, and at the same time exemplify ways of increasing their position within the marketplace.

- The dramatic decline of the farming population over the past century could mean that farmers regard themselves as left behind and out of step with modern social change, but how farmers interpret their choice of career and lifestyle could also encourage a different view of how farming has changed.

These are among the characteristics of contemporary farming that shape how farmers think and talk about farming. Many of these characteristics are ones that have been of interest in broader scholarly discussions as well. How family ties and business relations can function together is one example. What it means to be an independent person when in many ways that is not the case is another. Why technology is embraced that may erode deeply held values is yet another.[5]

American farm life is vastly diverse—far more diverse than the interchangeable bushels of wheat and gallons of milk that get tabulated in farm statistics. The commonalities that may appear from the fact that farmers live in the country and earn their living from the land are refracted through the different lenses of topography, soil, and location. Farm life varies with seasonal changes in the weather. It is quite different for someone managing a spread of ten thousand acres than for a family earning a living from fewer than a hundred.

The true diversity of farm life is evident in the meanings that farmers attach to it as they tell their stories. The land holds distinct meanings because it has been in the family for several generations. Or it has meaning because its value is increasing. Or both. That grove behind the barn, a farmer might say, is where I played hide-and-seek growing up. I hated getting up in the morning to help milk the cows. Somehow I just enjoy being out on the tractor and looking out across the field.

The way to gain an understanding of what farm life means, short of farming oneself, is to listen as farmers tell their stories—as they talk about what they like or do not like, why they went into farming and why they have stayed, how it affects their families and what happens when they talk with their neighbors, whether it somehow connects with their religious beliefs, how they think about the land, and how they feel about new technology and changes in the market. From these accounts it is possible to gain an understanding of the diverse ways in which farmers interpret their lives.

The fact that there are different sides to the story is important too. Neil Jorgensen's narrative about his years' farming offers a suggestive illustration.

Toward the end of a lengthy interview in which he spoke about the ups and downs of farming and what he does from day to day, he said that he was optimistic about the future of family farming. He would tell anyone considering it that it is a good life. Then, when asked if he wanted to add any other comments, he paused for a moment, seeming to hesitate, and said, yes, there was one thing. He talked for several minutes about the physical risks involved in farming. He described a serious accident that had put him in the hospital. But that was not as bad, he said, as an extended period of severe depression.

His depression might have manifested as seriously in any other line of work, but he was convinced that the struggles, the risks, and the uncertainties of farming made it worse. He eventually recovered. And yet, it was a struggle he wanted us to know about. "I prayed to die," he said. That's how bad I was."

This was one of many frank and personally revealing comments that emerged in our interviews. Farmers told of serious farm accidents and even murders that had taken place in their communities. They mentioned conflicts between husbands and wives and between parents and children. They talked of struggles over land and difficulties making ends meet. The stories were not told to show that farm life is terrible. Only that it is human.

Farming is inherently about families. The conclusion that came through clearly in our interviews is that farm families do work together, they do so across gender lines and often across generations, and these relationships are complicated by the fact that running a business and doing things as a family converge so often and in such complex and sometimes conflicting ways. As farm life changes, farmers argue that family relationships are still among their highest priorities. They enjoy working together and insist that farms are good places to raise children. And yet these family relationships are changing. Farmers are in the position of having to invent new reasons for arguing that farms are good places for families. I examine these reasons and their underlying relationships in chapter 1.

Farmers' relationships with neighbors are changing as well. The idea that farming communities are places in which neighbors understand one another, share work, worship together, drink coffee together on slow mornings, and enjoy one another's company is an ideal that many farmers would like to maintain. But they are finding it harder to realize this ideal in practice. Looking closely at what they do and say about neighbors suggests that neighborliness is being maintained in ways that depend less on warm feelings and more on formal organizations. The role of neighbors is the focus of chapter 2.

Like neighborliness, religious sentiments among farm families also appear to be changing. If sacred narratives about good shepherds and abundant harvests bear continuing resonance in farming communities, houses of worship are less often filled than in the past because of declining farm populations. Because of their enduring attachments to the land, farm families typically live in their communities for periods spanning lifetimes and generations. That lends stability to rural congregations. However, it can also breed discontent that may be especially difficult to transcend. How farmers experience faith and talk about it is discussed in chapter 3.

With farm life embedded so clearly in families and communities, an observer would have to wonder what farmers might say about being independent. That image of rugged, strong-willed independence has been part of legends about American farmers from the beginning. In these narratives farmers are the epitome of American individualism. The farmers we spoke with still embraced an ethos of self-determination. Being their own boss was what they especially liked about farming. They evaluated success and failure in these terms. At the same time the evidence suggests that the meaning of personal independence is changing. Chapter 4 summarizes what farmers said about their understandings of independence.

These shifting ways of understanding farm life ultimately bear on farmers' relationships to the land. On the one hand, the land is almost like family. It conjures up deep feelings of respect. Adoration sometimes borders on worship. On the other hand, farmers' relationships to the land are mediated by big machinery that reduces their immediate physical contact with the soil, by bank loans and soaring prices, and by chemicals. The resulting understanding is at best one of ambivalence. Farmers want to be good stewards of the land but express uncertainty about how best to practice good stewardship. Chapter 5 presents conclusions about farmers' understandings of the land.

The change that farmers say is affecting their lives most powerfully is technological innovation. Larger and more expensive machinery, genetically engineered seed, new fertilizers and pesticides, and information technology are all affecting farm life dramatically. Many of these developments are ones that farmers eagerly embrace. At the same time they are caught up short with questions they cannot answer about the best uses of technology and where it is all heading. How the farmers we spoke with think about technology is the focus of chapter 6.

The challenge that keeps farmers from sleeping at night—other than uncertainties about the weather—is concern about markets. They know that markets for farm commodities have never been under their control. But they worry that market fluctuations are occurring more rapidly and in larger swings than ever before. The fluctuations appear to be random and unpredictable and yet seem to be increasingly shaped by traders, by an agribusiness plutocracy, and by foreign countries. Against those odds, a striking number of the farmers we spoke with nevertheless described small ways in which they hoped to gain some control over the markets in which they function. Farmers' views of markets are discussed in chapter 7.

No single story emerges from these conversations. Nor should it. These are not the observations of policymakers who worry about food supplies and corporate agriculture. Nor are they the descriptions of lives left behind by those who have moved on to other places and different careers. They are the experiences and the meanings of those experiences of farmers who have stayed in family farming. They show what family farming is like and how it is changing. The message is that farming is complicated and yet inflected with family stories, relationships, and experiences drawn from day-to-day activities that render it uniquely meaningful to those involved. This is the message farmers we spoke with hoped the nonfarming public would understand.

FAMILIES

1

It's in your blood somehow. I don't know how to explain that. I think
there are a lot of people who are involved in agriculture in spite of the
fact that it's difficult to make a living, just because they can't get away
from it.

—Dairy farmer, female, age 57

Things have changed in farming, but the things that have not changed
are pride in family, raising a decent family, loving them. Those are my
ideas of success.

—Corn-belt farmer, male, age 64

The Jorgensens' intergenerational corn and soybean farm illustrates a pattern
that has been present in many farming communities for years. Although
Neil is in his forties and has been farming for more than two decades, his
parents are still actively involved. Living only a mile and a quarter away,
Clay and Mary pitch in during corn and soybean planting and in harvest
when extra hands are needed. Apart from Arlene's job in town, the two
families' income depends entirely on the land.

Whether they have formed an official partnership with some other
member of their family or whether they merely enlist the help of a spouse,
parent, child, or sibling, family relationships are the very core of family
farming. We wanted to hear what farmers had to say about these relationships. What relationships were involved? Who did what? How were these
relationships structured? What was fulfilling about them? What was trou-
blesome? And what were the social norms and values that these relationships
embodied?

When Americans talk about valuing their families, the remarks usually
emphasize the priority of investing resources in family life. These resources
are attained outside the family, such as earning income by working at a job,

or are ones that could have been invested elsewhere, such as time spent at home instead of at work. As entities in which resources are invested, families are in this sense units of consumption.

Farms are different. Although farm families are units of consumption, they are also units of production. The Jorgensens produce corn and soybeans. Other families grow wheat or cotton or sell milk and vegetables. Family life and farming are operationally intertwined. Home and business are integrally connected.

In the nineteenth century family involvement in farming was best illustrated in family members' labor and in the fact that most of these activities took place in close proximity to the farmstead. In the twenty-first century the locus of activity and the nature of those activities have changed. But families are still operationally involved and the meaning of farm life rests on that involvement.

The academic literature describing relationships between families and business activities has changed in ways that take account of the greater complexity of these relationships. Earlier arguments stressed the advantages of sharp differentiation between families and business activities. Lacking such differentiation, business activities could be hindered by unproductive family members. Differentiation allowed businesses to hire and reward only the most productive workers. Differentiation involved not only spatial separation but also contrasting legal codes and principles of valuation.[1]

Those arguments reflected scholars' impressions of large industrial enterprises, such as textile factories and steel mills, but underestimated the extent to which efficient business enterprises can also retain close relationships with families. More recent discussions acknowledge those relationships. Family ties can sustain business activities during difficult times and family relationships can be the means through which capital is accumulated and specialized knowledge is shared.[2]

The connection of farm families to farming includes real and symbolic linkages with the farming traditions and values of previous generations. It still involves contributions of labor from family members at least during peak seasons and often consists of complex formal and informal partnership arrangements. These relationships endure but are also being renegotiated as farming changes. Among the most important changes are shifts in gender roles and different modes of transmitting family values to farm children.[3]

FAMILY TRADITION

Herbert and Darla Loescher are dairy farmers. They get up at four o'clock every morning and do the chores together. One gets the seventy cows they own milked. The other feeds the calves and cleans the mangers. The Loeschers are in their mid-forties and have been doing this all their adult lives.

13

Longer, actually. Mrs. Loescher started helping with farm chores when she was eight. Mr. Loescher cannot remember when he started. He just grew up around cows.

The Loeschers live in a tight-knit farming community of German Americans who settled here in the 1870s. Mr. Loescher's grandparents ran a hundred-acre dairy farm and raised five children. Mrs. Loescher's grandparents had a two-hundred-acre dairy farm and raised twelve children. Mr. and Mrs. Loescher were both raised on dairy farms. As children they saw their grandparents almost every day and grew up doing farm chores. They feel strongly that they are following in their ancestors' footsteps by farming. "Once it is in your blood," Mr. Loescher says, "it stays in your blood."

The fact that many if not most farmers were raised in farm families is one of the most distinctive features of farm life. It is hard to imagine any other occupation in which this kind of generational continuity is as important. Even in families in which more than one generation has worked as teamsters or taught school, the chances of more than two generations having been involved and having worked at exactly the same location are lower in most instances than among farmers.

One way of thinking about intergenerational continuity in farming is that it is strictly a matter of economic considerations. Young people wanting to farm have an advantage if their parents and grandparents have farmed, own land that can be passed along, and have machinery to share. Those advantages are important, but they do not illuminate the meanings and values that get passed from generation to generation.

Being significantly connected to the past is as important to the mentality of farming as it is to the economics of farming. This experience of intergenerational continuity is what farmers mean when they say that farming is in their blood. They are following in their parents' footsteps—often in their grandparents' and great-grandparents' as well. They are carrying on a family tradition.[4]

Farm families for this reason cannot be understood in terms of relationships only among the living. It would be inaccurate, for example, merely to count whether farmers were married and how many children lived with them or nearby. Family farming means relating to the past as well as to the present. "The dead are all around," a third-generation farmer noted. "They are with us every day." He meant the parents and grandparents who preceded him farming. Their voices still echoed in his head.

The voices are words passed from fathers and mothers to sons and daughters. They include "little wisdoms," as one farmer put it, such as "live poor and die rich" or "lots of people are smarter than you" or "when there's work to be done, do it." Some of the words are stories. The narratives tell of memorable episodes. The barn caught fire. The young farmwoman whose family once lived down the road died in childbirth. More of the words are cues to

14

ways of being. They encapsulate the family tradition and convey ideas about the meanings of daily life.[5]

Tradition is commonly understood as a way in which authority in social relations gains legitimacy. Traditional authority is legitimate because of how things were done in the past. Knowing that a person's ancestors farmed in this place for several generations gives legitimacy to that farmer's choice of career and indeed to the farmer's sense of how and why things should be done in a particular way.

The dead hand of the past, as the saying goes, casts tradition in a different light. If having ancestors who did things a certain way provides legitimacy, veneration of the past is capable of stifling innovation as well. The fact that family traditions are an important part of farm families' understanding of themselves puts them at risk of either being reluctant to adapt to changing conditions or of being regarded that way.

But family traditions serve purposes other than legitimating the past. They serve especially as the mechanisms through which family members identify and affirm their membership. Traditions are stories told at the supper table, during family gatherings, and in letters and conversations (or as the case may be, in e-mails and on Facebook). Traditions in farm families are communicated in these ways as well as over coffee at the local co-op and while waiting for a spare cultivator part to arrive. They show the relationships among family members. They demonstrate that insiders are better able to understand the stories than outsiders are.

In real life the stories of which family traditions are composed have several prominent themes. Many of the stories are narratives of origin. They demarcate history from prehistory. "See that mound over beyond the barn," a woman who lives on the small truck farm her husband inherited from his parents says. "That's where the Indians used to bury their dead." It reminds her that people were here long before anyone in her family. But she also knows which of her husband's ancestors were the first to farm in the area and when that was.

Sometimes being associated with tangible objects enriches the stories. When families have farmed in the same location for several generations, the stories relate to the old well where the windmill used to be, the rusted horse-drawn plow that has been preserved as an antique, or the butter churn that came with great-grandparents by covered wagon. "The dinner bell out there in the front yard," a woman in the cotton belt explains. "I still ring it on special occasions. It was there when this was a cotton plantation with slaves."

Whether the stories describe mundane farming activities, recount tragedies, or convey humor, the narratives define something special about farm life. They draw distinctions between persons who understand the special meaning of the stories and persons who do not. To truly appreciate the tale, the audience has to share some of the speaker's background knowledge.

15

An example is a lengthy story a farmer in his sixties told about his grandparents. The story described a Sunday dinner during which his grandparents hosted a college professor they somehow knew and who had little knowledge of farm life. Part of the tale included the cloth convertible top on the professor's car being eaten by goats. Another part involved the professor being told the chicken they were having for dinner had drowned in the horse trough.

The farmer telling the story knew his retelling it was part of an interview for a professor. He also understood that stories like this played an important role in perpetuating the distinctive family meanings he associated with farming. "These stories are very valuable," he said, "because they pretty much define the culture and they define our family values."

Family traditions are mental sinews connecting past and present. Farmers' identities bear specific continuities with previous generations. Their parents not only farmed, they farmed *here*, in this place, and they raised corn in that field and milked cows in that barn.

Or at least they farmed in the same community. Subsequent generations know where the home place was located. The old barn may have crumbled years ago. The trees may be gone. But in their mind's eye the current generation can visualize the old buildings. Fields are known by their previous owners. That was Joe's place. Over there is where the Dubrovniks lived.

Neil Jorgensen's sense of farming being in his blood is closely associated with having lived in the same place all his life. "It's a pride thing," he says. "You take care of the land and the soil and try to keep it from generation to generation." He values keeping the family's name attached to the land.

The Loeschers have been less fortunate. Although they live in the community in which they were raised, they do not live on a farm that has been in the family for generations. The land they now farm is rented. Their dream is to become landowners like their grandparents.

Traditions rooted in particular spatial connections are strengthened by continuities in the skills involved. Although farming changes from generation to generation, there are similarities as well. Farmers talk about doing things as children that they still do as adults. They often mention helping to feed and care for animals in this regard. The Loeschers have been getting up early to tend the cows for as long as they can remember.

"It was nothing to be up at six o'clock to go out and do hog chores before school," Neil Jorgensen remembers. He started doing this in third grade. Up until a couple of years ago he was still raising hogs. "I've been with hogs all my life," he chuckles.

Clay Jorgensen illustrates another aspect of farmers' relationship with the past. Family connections made it both possible and plausible to farm.

He started a year before graduating from high school. His father had died. His mother and brother were keeping the farm going but were struggling. His mother told him she had a tractor and a few implements. He could have them if he wanted to farm. Or she could sell them and he could go to college. College was an unknown world. He had no idea what he could do with a college degree. He opted to farm.

For Mr. Jorgensen farming was a risky choice because his mother had barely enough land for him to eke out a living. But farming was more plausible than any conceivable alternative because there was more to family than simply the idea of farming. His grandfather and two great-uncles were still farming in the neighborhood. He could share equipment with them and call on them for advice whenever he needed it. "If I called on them to help," he said, "they'd come over for a day. We'd make hay and shell corn and maybe do a little of the harvesting together."

Traditions like these were more commonly passed among the farmers we spoke with from fathers to sons than from fathers or mothers to daughters. Girls grew up doing farm chores like boys did. And mothers played important roles, as Mr. Jorgensen's mother did, in keeping land in the family when they outlived their husbands. But farmers we spoke with were more likely to have had parents who encouraged them to follow in their footsteps if they were boys than if they were girls.

The Loeschers are typical in this regard. Mr. Loescher says his parents would have been happy if he had chosen some other line of work, but they clearly hoped he would farm. Seeing his mother working alongside his father on the farm pushed him in that direction. Mrs. Loescher's parents did not encourage her to farm. Her mother had a job in town. Mrs. Loescher grew up thinking she would marry someone who did not farm. She surprised herself by falling in love with a farmer.

One of the instances we found in which the family tradition stemmed more from the wife than from the husband was Janelle and Michael Bower. The Bowers are a couple in their early forties who farm in wheat-growing country. They are fourth-generation farmers who operate a medium-sized farm of about fifteen hundred acres. Mrs. Bower's great-grandfather had been one of the area's first settlers back in the 1870s.

Most of the land she and her husband currently farm was owned by her mother and had been farmed by her father and grandfather. Mr. Bower had grown up in another state and had some farming in his background, but his father had a job in town and only farmed a few acres as a hobby. Mr. Bower imagined he would become a teacher when he graduated from college.

When the Bowers got married, it was her father who persuaded them to farm. He wanted someone to farm with him, did not have a son who was interested in farming at the time, and hoped his daughter would live nearby.

17

Mr. Bower and his father-in-law farmed in partnership for more than a decade before the older man retired.

As these examples suggest, family tradition exercises a strong pull in farming communities. Farming the same land as one's ancestors is a way of honoring the tradition. The spatial connection is not only with the land, as one woman especially wanted to emphasize, but with a "whole lifestyle" that she said was hard to describe but that she deeply wanted to protect. The connection is not even in doing the same kind of work as one's ancestors did. The work may be quite different. It is rather that farming itself is valued and that value is shared within families and across generations.

These generational continuities are so common that one might wonder if there are exceptions. Is it possible for anyone to farm who has not been raised farming? Or has American farming become an entirely inherited way of life?

Mort Lancaster is one of several farmers we spoke with who did not grow up farming. A thoughtful and articulate farmer in his fifties, he illustrates the possibility of becoming a first-generation farmer, but also the special circumstances that may be involved.

By the time he graduated from high school in the 1970s the wheat belt where Mr. Lancaster lives was already populated with third-generation farmers who typically owned or rented more than a thousand acres. His family lived in town, and his father worked for a large oil company. Nobody who knew him then would have predicted that the Lancaster boy would become a farmer, let alone one of the most successful farmers in the area.

After working as a truck driver for several years out of high school, Mr. Lancaster decided he would rather do something more varied and interesting—something where he could be his own boss—even if it meant less take-home income. "I really wondered if I could do this," he recalls, because land in his area was usually kept within families. Still, he knew something about farming from having worked for a farmer in the area during junior high and high school. He rented a pasture, took out a loan, and purchased some cows. One thing led to another. He lived frugally, postponed marrying and having children, and bought used machinery. A decade later he had enough income and enough rented land to start investing in better machinery and land of his own.

But this is a story about family connections as well. Mr. Lancaster's father did not farm, but his grandfather did. When his grandfather died, Mr. Lancaster's father inherited several hundred acres of land, which became available for Mr. Lancaster to farm. Mr. Lancaster's wife, Sara, also brought valuable experience to the farm. Although she was raised in another part of the country, she grew up on a farm, knew the cattle business first-hand, and had a college degree in farm management.

It is also a story about timing. Mr. Lancaster says he was fortunate to have started farming in the 1970s, when interest rates were low. He could not have done it a few years later, when interest rates were high. He benefited from land becoming available in the 1980s as marginal farmers went under. He hopes one of his sons will be able to follow in his footsteps.

HARD TIMES

Carrying on the family tradition is all the more special, farmers say, because of the hardships earlier generations faced. Nearly everyone we spoke with had stories to tell of difficult times. Fourth- and fifth-generation farmers talked about their ancestors coming to the area as pioneers, breaking the sod for the first time, facing danger, and losing crops to grasshopper invasions. They were proud to be carrying on the legacy of these hardy predecessors.

As an example, consider how Stan and Elizabeth Rayburn describe their heritage in farming. They are cotton-belt farmers who are doing well by local standards. Their modern ranch-style home sits well back from the road. Two giant machine sheds dwarf it in size. They farm three thousand acres in partnership with Mr. Rayburn's brother. Part of the land—a quarter section of flat, treeless soil—has been in the family since the 1880s. It means a lot to Mr. Rayburn because his grandfather and his grandfather's mother acquired it. They were sharecroppers eager to have land of their own. He knows their story by heart and has retold it many times.

When unbroken land hundreds of miles to the west of where they originally farmed became available, they loaded their belongings in a wagon, hitched it to a team of horses, and walked a cow or two alongside. They made trip after trip, going at it six days a week. "I was told it took them about three and a half months," Mr. Rayburn says. He doesn't recall all the details but remembers hearing the story growing up whenever relatives got together.

Third-generation farmers we spoke with generally had memories of parents and grandparents suffering during the Great Depression. They knew of struggles to pay off mortgages and hold on to the land. Their grandfather broke his leg and couldn't work. Their grandmother taught at a one-room school. She made enough to put food on the table. They endured extreme hardship and were fortunate to avoid losing the farm.

There were family memories of dust storms, drought, and bank failures. People saved what money they could, never threw anything away, and taught their children to want little. Neighbors gave up and moved away, became day laborers to put bread on the table, or quit farming entirely.

Neil Jorgensen knew his grandmother well when he was growing up. She was a widow by that time, a meticulous woman who liked to keep

19

everything well organized and took a daily interest in the farm. During the Depression she and her husband lost all their savings when the local bank failed. For the rest of her life she never quite trusted the banks. She banked at several different ones to make sure the Federal Deposit Insurance Corporation (FDIC) insured her savings. "She was very cautious," Mr. Jorgensen says. "Didn't spend anything." Images of her as an elderly woman hoeing soybeans and cutting thistles in the pasture are still clear in his mind. He takes pride in following her example.

A hardship story is one thing if someone merely learns in school that times were difficult for Americans who lived through the Depression. It becomes more meaningful if those people were one's grandparents. Farmers' accounts bring the stories even closer. When grandparents were a daily physical presence in their grandchildren's lives, the stories are told with greater regularity and with the added significance of familiar places, objects, visual reminders, sounds, and smells.

A third-generation wheat-belt farmer points to the barn where his grandfather taught him to milk cows and told him about the time they almost lost the farm. Near the barn is the chicken house. His grandmother used to go there when the dust storms came and throw the chickens across the room to keep them from piling up and suffocating the ones on the bottom. A corn-belt farmer sits at the kitchen table where his grandmother cooked meals while his parents tended the crops. The Depression taught her to be frugal. She taught him to always clean his plate. He feels a visceral connection with the hard times she experienced.

Hardship during the 1930s was long enough ago that farmers might have forgotten its lessons except for the fact that there were also more recent crises and difficulties. Clay Jorgensen says he almost went under several times. As a young farmer, he got hailed out the very first year and then the second year he harvested almost nothing because of the drought. In the 1980s interest rates almost killed him. The rates rose to 18 percent one year. He considers himself fortunate to have survived. Many of his neighbors did not. His son has heard the stories many times. Neil can give the exact years when each of these events happened. The lesson is that farming takes fortitude. It requires just "plugging along."

A neighbor of Mr. Rayburn who also farms cotton is a more recent farmer in that area. His stories of hardship are situated in the 1950s. His father was doing ditch irrigation at the time, which required constant monitoring. With no hired help, the man's father would go out and lie in the field, put his arms out, and sleep. When he felt the water touch him, he would wake up and know that it was time to change the rows. This was in an area where there were rattlesnakes.

There were other stories of farm injuries, houses burning down, and relatives dying. Some of the memories were no different from anyone recalling a parent or grandparent who had died, perhaps young or tragically. The difference for farmers was that the stories were woven into the meaning of farming and of carrying on a family tradition. The person who died was the ancestor from whom the current farm was inherited. Or the farm accident was the kind that still posed dangers.

An example was an account given by a farmer in his sixties of his wife's father who died in a farm accident while the children were still young. The man was killed while using a mechanical hay stacker. Besides the fact that it was a tragedy for his wife's family, the man telling the story considered it especially pertinent because he himself did similar work and now farmed some of the same land.

The story was part of a larger narrative about the dangers of farming. Another relative died young of a heart attack. Yet another was having trouble making a living farming so worked part-time at a nearby seed and fertilizer plant. That relative died from what the family suspected was chemical poisoning.

The stories of more recent difficulties focused on seasons when bad weather wreaked havoc. The land where Mr. Rayburn grows cotton is mostly irrigated now, but drought is still a problem. Just this past year the drought was one for the record books, he says. The old-timers there compare it with earlier droughts. Mr. Rayburn's mother still lives just down the road. She says the recent drought is nothing like the dry years she experienced in the Dirty Thirties. There were bad years in the 1950s, though. Farmers tell how they suffered and what they did to survive. The stories convey more clearly what farmers' experienced than average rainfall statistics do.

Other than bad weather the most common source of misfortune was bad decisions about market transactions. Some of these decisions had long-term consequences. Clay Jorgensen, for example, recalled his decision years ago to keep his cattle instead of selling them even though the market was plummeting. He lost so much money that year that he was unable to make the payments on land he had recently purchased from his own income and had to borrow money from the bank. He has had to borrow money ever since. "The banker should take me out to dinner sometime," he muses. Mr. Jorgensen was able to pay for the land but never got back on his feet enough to buy more.

The stories of hard times seldom implied that farming was a bad life. Only that it was difficult—that there were seasons when crops failed, years when earnings were low, and days when illness or accidents made hard physical work all the more challenging. And that meant taking some pride in one's family having survived. Indeed, the moral of the story was usually that it took grit to survive. As one farmer put it, "Yes, there were hard times, but they made it go." The rest of the typical story was that people hunkered down, weathered the drought, shifted to a new crop, and held things together.

21

COULD'VE BEEN WORSE

With all their stories of hard times, what keeps farmers going? We asked the farmers we spoke with if they had ever seriously considered quitting. Many of them had, but only momentarily. After the drought left them with no crop or a hail storm ruined the harvest, they admit thinking it would be better to do something else. They had relatives and neighbors who had in fact quit. That awareness deeply shaped their understanding of what it meant to still be farming.

Mr. and Mrs. Loescher laugh when asked if they ever considered quitting. "Plenty of times," she replies. A few years ago they thought seriously about quitting. They lost the land they were renting and had to move. Milk prices were low, hardly covering expenses. Several of their neighbors sold out and started doing something else.

In reality getting out of farming would probably have been easier for some of the farmers we spoke with financially than psychologically. In these instances the land could have been sold at a handsome profit. The psychological cost, though, would have involved abandoning a family tradition.

"A family farm is history," Mrs. Loescher explains. "It is love; it is caring." She says it is not at all like working somewhere just for a paycheck or living in a place where you are nobody. "Just the word 'family' says it all. It is family. You care."

The argument we heard most often about why they stayed in farming despite hard times was that things could have been worse. Milk prices were low but not as low as they could have been. A job in the city might have paid better but would have been bad for the family. Selling the farm would be like betraying one's ancestors.

Just as there were stories about hard times in farming, there were cautionary tales about relatives who left the farm and regretted it—or should have. One example was a story of relatives who moved to the city and just hated living so close to their neighbors. Another tale featured an uncle who went off to college, landed a big corporate job in the city, and then became an alcoholic.

While Mr. Loescher was growing up, his father worked part-time at a factory in the area doing manual labor to help make up for the farm's meager earnings. Mrs. Loescher's father did the same thing. Mr. and Mrs. Loescher have their fathers' experiences as a point of reference. When earnings are low and the work is hard they still think things are relatively good.

Farmers like the Loeschers acknowledge that they had few other options. They could have worked as hired hands for other farmers. Some worked part-time in construction and knew that would be their most likely alternative to farming. Others knew their best alternatives would have been driving a truck, working as an oil company roustabout, or pumping gas. They also told stories meant to show that even though farming had its ups and downs, other lines of work were even worse.

"They just fell out of bed!" This was Clay Jorgensen's way of talking about lines of work that were worse than farming. He worked part-time as a young man for a farm implement dealer. He enjoyed the mechanical labor involved and the opportunity to talk with customers. He thought he might even enjoy selling farm implements more than farming. But he is glad he did not pursue that idea. During the 1980s, when farmers were having difficulties, the implement dealers in his area mostly went out of business entirely. "They just fell out of bed," he observed.

It would be wrong to conclude that most farmers had few attractive alternative occupational choices, though. Among the growing number of farmers with college educations, farming was one of several possible alternatives. Many of the farmers we spoke with had considered other careers during college. Some worked at other jobs while they began farming on the side and earned enough from their off-farm job to gradually purchase land and machinery.[6]

Farmers who had worked in other middle-class jobs less often described why it could have been worse staying in these alternative occupations in economic terms. Not farming in these instances meant doing less interesting work, having to work for someone else, or having to live in a city. It implied being unable to pursue one's dreams.

The notion that things could have been worse extended beyond comparisons with alternative occupations. It was a kind of idiom that farmers used to express a broader outlook on life. Sure, the drought was bad this year, but it could have been worse. Income barely covered expenses, but things could have been worse.

Neil Jorgensen illustrated this outlook while talking about the financial risks farmers in his area were taking. "It used to be being cautious," he explained in describing how farmers dealt with bad years. "But now you've got to take that risk or you'll be left behind." That reminded him of a comment he'd heard recently from a neighbor. The neighbor was heavily in debt to the bank and was discussing a recent crop failure. "It could have been worse," the neighbor chuckled. "It could have been my money instead of the banker's money."

The farmers we spoke with were quick to acknowledge, too, that the stories they heard from previous generations were not all bad. The reason their father farmed, they said, was that he really enjoyed raising cattle or hogs or loved being outdoors or working with horses or driving a tractor. Especially when they grew up seeing a parent enjoy farming, they understood this to be part of the story. They knew firsthand how a father or grandfather had been reluctant to retire from farming. Having farming in one's blood meant sharing some of those sources of enjoyment.

FAMILY BUSINESS

Family farming of course is more than simply following in the footsteps of previous generations. It involves business relationships as well. The principal business relationships that the farmers we studied described were husband-wife operations, nuclear-family-plus arrangements, informal partnerships, formal partnerships, and family corporations.[7]

Although husband-and-wife operations were the most common arrangements among the farmers we spoke with, many of the farmers were involved in various informal or formal father-son partnerships like the Jorgensens. Other arrangements included father-daughter and son-in-law partnerships, brother-brother partnerships, and family corporations involving aunts, uncles, and cousins. Some of these were formalized legal relationships of the kind that show up in USDA statistics as partnerships and corporations. Those formal relationships, though, capture only a small part of the larger picture. The farmers we spoke with had worked out a wide range of agreements with family members that were integral to their farming operations.

The simplest arrangements are farms operated solely by husbands and wives. Nearly all of the farmers we spoke with were married. In these cases the relevant division of labor depended mostly on whether the wife or husband or both held off-farm jobs. Usually the husband did most of the fieldwork, the wife ran errands and operated equipment during harvest or planting, and one spouse or the other or both handled the bookkeeping and paperwork.

What might be described as nuclear family operations in most cases were more accurately characterized as nuclear-family-plus activities. "Plus" meant that additional labor was usually involved, at least during peak seasons. Adult children, siblings, and cousins helped during planting and harvest. Retired parents provided assistance. Perhaps a part-time hired hand was available as well. It helped that retired relatives lived nearby and that relatives with town jobs did so as well.

Nuclear-family-plus arrangements were especially important among the farmers we spoke with whose work involved labor-intensive activity during peak seasons. Examples included harvesting wheat and corn, picking cotton, planting fall crops, cutting ensilage, and birthing calves. Despite larger and better equipment that reduced the need for manual labor, the fact that peak demand for labor occurred for all neighboring farmers at nearly the same time increased the importance of being able to secure additional labor through family networks.

Dwayne and Carol Hebner are an interesting case in point. They are wheat-belt farmers in their early forties who annually plant approximately 1,600 of their 2,700 acres to wheat. With a 30-foot header on the almost-new combine the Hebners recently purchased for nearly a quarter of a million dollars, they can harvest at least 150 acres a day. A typical harvest day begins at 5:30 AM with fueling and greasing the equipment and ends around 11:00 PM. Mr. Hebner runs the combine. Mrs. Hebner packs lunches, brings supper to the field, helps with the grain cart, and runs errands when parts and supplies are needed. His parents live nearby. His father drives the loaded semi to the elevator in town. His mother helps with the meals. Her parents are retired and live in a different part of the state. They come each harvest and stay for the duration. They also help with the equipment and meals.

Husband-wife farms that did not involve actual partnerships included business arrangements that crossed generations. Financial help from parents when farmers were starting out was nearly universal. It consisted of parents renting land to sons or daughters who otherwise may have had difficulty obtaining land. It included parents co-signing on bank loans. Other forms of assistance ranged from providing the barn and equipment for fledgling dairy and livestock operations, to sharing machinery, to cash loans, to exchanges of labor for machinery or land. These arrangements were a significant contribution to the opportunities for younger farmers to follow in their parents' footsteps.[8]

Younger farmers benefited mostly from parents but also from grandparents, aunts and uncles, cousins, and siblings. These extended family networks were frequently the source of land that became available to purchase or rent and stayed within the extended family because the owners wanted to carry on a family tradition. One farmer was able to get into the cattle business because an uncle owned a feedlot. Another worked for an uncle while starting to farm to help make ends meet and learn the business. Others spoke about sharing equipment.

Besides financial assistance, husband-wife farms frequently benefited from having parents nearby to consult with about business decisions. "There's just a lot of paperwork that needs to be done," a farmer in his sixties who is helping his daughter and son-in-law start farming observed. He thinks it helps living just down the road. "If we were like fifty miles away and they're trying to describe something over the phone, I couldn't understand it. But I can see the paperwork." He feels this kind of assistance is even more important than when he started farming.

Actual business partnerships among the farmers we interviewed ranged from formalized arrangements of the kind reported in agricultural statistics to informal agreements involving working together, sharing machinery, and exchanging labor. The most common partnerships were between fathers and sons and brothers. We also talked with farmers who worked in partnership with daughters, sons-in-law, nephews, uncles, and cousins. Father-son partnerships ranged from ones in which the son was just getting started and needed the father's financial assistance to ones in which the son was well established and the father was still involved despite being old enough to have retired.

The partnerships were nearly always with relatives rather than with unrelated neighbors and friends. Partnerships with blood relatives were more common than with in-laws. Usually the partnership reflected patterns of inherited land and the division of land within families across generations. Each partnership involved distinctive understandings about the sharing of land, machinery, labor, and income.

The complexity of these arrangements is evident in the Jorgensens' father-son partnership. Clay and Neil describe the partnership simply as a 50-50 arrangement, meaning that they share equally in revenue from the corn and soybeans they sell. However, they have worked out a number of informal agreements over the years that are intended to equalize the labor, capital investment, and other expenditures involved in what they share and to set aside other aspects of the business that they do not share. For instance, each owns a tractor separately from the other, the combine and most of the land is owned by Clay, Neil does the lion's share of the physical labor and rents some land apart from his father's, and Neil owns cattle separately but also helps Clay with his. To further complicate the relationship, Clay has helped Neil at times by providing additional income from crops and through assistance in getting started raising hogs.

Mr. Rayburn's partnership with his brother is similar to the Jorgensens' father-son arrangement. Each brother files separate personal income tax reports and has a separate family budget, but the paperwork for the land they farm together is filed on behalf of the partnership. They do not keep track precisely of the hours each spends on the paperwork or for that matter on fieldwork and other farm activities. Having grown up together and farmed together all of their adult lives, they operate mostly on trust. "We pretty much know how the other one is going to want to do things," Mr. Rayburn says. During planting time and harvest they work side by side in the field. At other times they work at different things, each making a distinctive contribution. "We kind of know where our specialties are in the operation," Mr. Rayburn explains.

Flexible arrangements based on trust and intermittent renegotiations sometimes falter and result in disagreements. This is the reason that other farmers we spoke with described relationships involving specific verbal and written agreements about hours contributed, machinery purchased, and expenses charged for machinery depreciation. Although they were related and knew each other intimately, they considered it important to avoid conflicts from misunderstandings. They worked out hourly wage agreements based on market rates and wrote up contracts, sometimes consulting financial advisors and lawyers in the process.

Tom and Allen Granger and their cousin Linda Farnsworth operate one of the largest truck farms in their state. With several thousand acres to manage, they spend more time managing than they do on fieldwork. The farm started with their grandparents nearly a century ago and grew significantly in the 1960s under the leadership of the Grangers' father and his brother, Mrs. Farnsworth's dad. The two men farmed in partnership, formed a family corporation called Granger Farms, and passed the land and control of the corporation to the current generation in the 1980s.

Nearly everything about Granger Farms is formalized as written agreements rather than relying on implicit understandings. The various parcels of land are owned by the corporation or jointly by siblings or by husbands and wives and thus are reported separately for IRS and USDA purposes. Additional rented land is also reported separately. Approximately thirty full-time employees work for Granger Farms. During and after college Tom and Allen worked as employees too.

These contractual relationships are necessary when so many family members and employees are involved. They lower the chances of misunderstandings. At the same time informal family relationships affect how major decisions are made. Granger Farms has foregone purchasing additional land and experimenting with new farming practices because of how those decisions would have affected family relationships. They point to conflicts among the owners of another family corporation in their community as an example of what they know could happen.

FAMILY CONFLICTS

Getting along in families is seldom easy. Misunderstandings, hurt feelings, and resentment come with the territory, no matter how well family members get along most of the time. When farm business relationships are involved, tensions are no less difficult to resolve. Although the fact that family members do work together may be an indication that they have to get along and do get along, money and work can aggravate the rivalries that may exist among siblings, with parents and children, and between spouses.[9]

"I guess everybody wants to be the Waltons," one woman observed, referring to the popular 1970s television series depicting an unbelievably warm and caring farm family during the Depression. But she added, "I don't know anybody who doesn't have some relatives they wish they didn't have."

Clay Jorgensen and his brother were still in school when their father died. Their mother and grandparents ran the farm until they graduated. Clay's brother was older. When Clay finished high school he started farming with his brother. For several years Clay did all the work while his brother was in the military. After that they continued farming together, sharing the land owned by their mother and sharing the work involved in raising hogs and tending cattle. But there were the usual rivalries between brothers, and those rivalries created significant problems between the two families. Eventually they decided the better part of wisdom was to break up the partnership.

The father-son partnership between Clay and Neil has seen its share of conflict, too. Neil's older brother decided early on that he did not want to farm with his father, and Clay agreed with that decision. Neil did want to farm, but he also wanted to be able to feel that he was capable of making his own decisions. It was one thing to have his father tell him what to do when Neil was twenty. That was harder when Neil was thirty and even harder when he was forty. On several occasions these control issues became so difficult that Neil thought about quitting. It took some serious rethinking and some outside professional advice to work through the conflicts and negotiate some clearer agreements about who is responsible for what.

Other farmers we spoke with mentioned similar conflicts associated with multigenerational farming. A frequent source of conflict involved the older farmer telling the younger farmer exactly how things should be done, rather than sharing the responsibility for important decisions. For example, a wheat-belt farmer recalled his father insisting that they each purchase identical tractors and identical cultivators even though different-sized machinery may have been more efficient. The younger man appreciated learning about agronomy from his father but recalled, "He had definite opinions. He taught me his way of farming, and one of the challenges is that things need to change and I'm trying to figure out what to change."

A dairy farmer and his wife recalled how the man's father objected when they decided to install milking machines. "His dad swore up and down that we were going to wreck all the cows because of those milking machines," the woman said. "He really wanted to hang on to the past." She can laugh about it now because that was forty years ago, but at the time it was difficult.

Another dairy farmer said he could have farmed in partnership with his father-in-law but passed on the opportunity in order to avoid potential conflicts. "I knew it wouldn't work," he said. "His personality and mine were too different." He especially saw things he knew he would want to do differently and knew his father-in-law would resist.

No matter how well matched father-and-son or father-in-law and son-in-law teams were in terms of personality, there were structural features of the relationships that led to conflict. One was that farming practices did change, and younger farmers were typically more informed about the changes and more interested in experimenting with them than older farmers. A second was that older farmers sometimes refused to retire long after their offspring were established, staying involved even despite declining health because farming was their life. A third was that fathers typically owned more of the land and machinery than their sons and thus had more at risk and felt it was their place to make the important decisions. A fourth was that debt had different implications. As the wheat-belt farmer explained, "We should have spent more and expanded more and had more debt." That was the way to expand early in one's career, he felt. But his father resisted. "When you're ready to retire, you don't get [into] debt."

Besides these farming-specific sources of conflict, there were generational differences that farmers sometimes associated with moral judgments. The older partners doubted that the younger ones, even their sons and daughters, knew how to work as hard as they had when they were young. Or they said the younger generation was materialistic or less careful about handling money. When moral judgments were absent, there was still a sense among the older generation that younger family members lacked the maturity to make good business decisions.

"I wish my son was half as organized as I am," a farmer in his sixties whose son was in his twenties complained. The son lived just down the road, and the father was getting him started farming. "He's a bit of a loose cannon," the father added. "He has a lot to learn yet. I just want him to get a little better grasp of things before I pull out or kick the bucket."

A few of the conflicts farmers described resulted in serious disputes, such as fathers refusing to help sons get started in farming, fathers retiring early or doing something else for a living, or sons quitting farming or severing a partnership. One woman related a story about one of her ancestors murdering his father-in-law. A farmer in another community said a couple of his neighbors were brothers who farmed together and one of them got mad and killed the other. "You just get to stewing on things," the farmer said, and then something like that happens.

But most of the cases we heard about farmers employed one of two strategies for resolving family conflicts. One strategy emphasized trust. The parties involved built on the fact that they were related, knew each other extremely well, and had each other's best interests at heart. They either talked a lot from day to day about decisions or held such similar opinions that they seldom needed to talk at all. These relationships included flexibility and allowed for adjustments to occur in routine activities because of also communicating in nonbusiness settings, such as having Sunday dinners together or visiting on rainy days. This was the kind of trust-building relationship that the social science literature has emphasized. Trust depended on close familial relationships and reduced the need to spell out business decisions in specific detail.

The other conflict-reducing pattern was just the opposite. It involved drawing a sharp distinction between familial activities, such as eating meals together, and business activities. The familial activities were congenial and informal. The business activities were spelled out formally in the same way that non-family-related partners or employers and employees would have arranged their relationships.

Mr. Bower described the best way of avoiding family conflicts, in his view, as involving "three spheres" that needed to be clearly understood in any family-owned business. The three spheres were family, ownership, and business. Family involved the kinds of issues that had nothing to do with business, such as deciding where and when to have Thanksgiving dinner. Ownership was the sphere governed by who owns the land and machinery. Business was "the daily management stuff." He thought conflict generally occurred when any two of those spheres were fused.

In Mr. Bower's case, his father-in-law was good at keeping the spheres separate. The two families could celebrate holidays together without discussing business. The ownership issue was readily resolved because it was customary in their area for whoever owned the land to receive a third of the proceeds and whoever did the work to receive the other two-thirds. The business part was hardest, but the two men tried to do equal amounts of the work, and because his father-in-law owned more of the machinery, Mr. Bower received 40 percent of the income and his father-in-law received 60 percent. In addition Mrs. Bower received hourly wages whenever she contributed farm labor and the two couples worked out an arrangement for health insurance coverage.

Whether they kept the spheres separate or not, other farmers we spoke with described family-business relationships that involved carefully prescribed understandings that were as formal as written contracts would have been—and indeed sometimes were written contracts. An older farmer, for example, might rent his machinery to a son or son-in-law for a specified dollar amount. A younger farmer might work for his father or father-in-law (or uncle or grandparent) for a specified per-hour amount. Other farmers described agreements about how much it was worth to live in a house owned by their parents or how they and a relative had split up the cattle herd or shared the labor of raising hogs.

One of the truck farmers had given a lot of thought to this issue. She said the downfall of family farms is often the older generation being unwilling to let go. She knows younger farmers in her community who are reasonably successful and yet whose fathers come around every day to tell them what they are doing wrong. She thinks the solution is a "business-like protocol" that spells out the relationship among family members in a contractual agreement. She believes it is especially important to have a transition agreement in place that prescribes when and how the farming operation will pass to the younger generation.

A well-known sociological argument contends that contractual relationships involving money are fundamentally at odds with personal family-style relationships involving intimacy. The one emphasizes specific utilitarian ends.

The other features diffuse enduring affective bonds. The differences are sufficiently stark that people are often reluctant to talk about money or conduct
business with family and friends for fear of conflict arising. This was a familiar concept among the farmers we interviewed. As one remarked, "If there's
money involved, there's always conflict."[10]

The interesting exceptions that sociologists have studied in which money
and intimate personal relationships comingle include prostitution, payments to surrogate mothers, the valuation of lives and limbs in personal
injury litigation, and clandestine markets for human organs. Family farming is another example. Although enduring intimate kin relationships are
present, the business aspects involve monetary arrangements.

The farmers we spoke with generally regarded clearly specified contractual relationships with family members as a way to reduce the chances of
conflicts occurring. Contracts specified which aspects of the personal relationships involved were to be treated as business arrangements and which
were not. Entering into a contract served as an occasion for the parties involved to state their understanding of what was fair.

The difficulty in negotiating contractual relations with family members,
farmers said, was that kin expected to be given preference over nonkin.
Or in other situations one family member was given preference over another family member who felt equally deserving and thus wronged by the
decision.

A wheat-belt farmer described a conflict that had arisen with a cousin.
The cousin had land to rent but rented it to someone else who was not a
relative. The farmer attributed the cousin's decision to jealousy. A farmer in
the same community mentioned a brother and sister who never spoke to
each other because one had refused to rent land to the other's son.

To avoid these difficulties, a community leader who had worked with
farmers for many years suggested, farmers needed to consider whether a son
or daughter or sibling was bringing assets to the business, rather than simply including them because they were kin. Were they bringing mechanical
skills or ideas about crops and markets that they learned in college or from
working in a different community? But doing that was hard, the leader
acknowledged. It could be misunderstood as being hard-nosed and money-
minded rather than demonstrating love.

A majority of our interviewees denied that conflicts of these kinds were
common among the people they knew. Several explained that few of the
farmers in their community were related to other farmers or landowners.
Most argued that people tried to be nice to one another. It was in their best
interest, they said, to avoid conflict with people you had to deal with the
rest of your life.

Attestations of harmony notwithstanding, nearly everyone knew of particular conflicts among their neighbors or within their own families. The

prevalence of these stories illustrated an important aspect of local culture. Word got around. Conflicts were known and discussed. They also served as cautionary tales. They reminded the storytellers and their listeners that family relationships could sour.

"We have brothers who farm around me," a truck farmer noted, "and it's just an interesting dynamic. They just go two different directions. I've watched it all my life. They help each other and I'm sure they've got each other's back. But one of them isn't afraid to steal the other one's water."

A corn-belt farmer had neighbors with a similar problem. In this case the brothers were such bitter rivals that neither wanted the other to purchase a farm that was for sale. They bid against each other until the one who got it paid almost triple the going rate.

Another corn-belt farmer said two brothers farmed just to the east of his place. Their mother sold her land to one. "He bought the land at a very, very good discount, according to what I see," he noted. "It wasn't fair."

"I've heard all sorts of stories and I don't know if they're true or not," a wheat-grower mused. He felt fortunate to have avoided conflicts thus far but knew they could happen. "There's no way I'm going to get my head caught between two people," he said, "because it could get squished pretty quick."

The lessons farmers drew from stories of conflict varied. Some were glad they did not have relatives nearby. As one noted, "If we were right next to each other, we probably wouldn't get along, but they're far enough away that we don't get in each other's hair." Some said conflict was just inevitable: farmers were stubborn by nature, siblings were born fighting. Another lesson was to dissolve family partnerships that got in the way of efficient business practices. Yet another was to formalize business arrangements to avoid potential conflicts.

Even the most carefully arranged business relationships were strained, the stories conveyed, when one of the parties died. This was the source of family conflict that farmers mentioned most often. In one community rumor had it that a farmer died and for two decades the heirs fought over the division of the estate. In another instance the son who took over the farm when his father died reportedly refused to divide the land or share the proceeds with his siblings.

Yet another incident involved a widow refusing to produce the necessary ownership papers to facilitate the transition to the next generation. In other cases the conflict began in anticipation of someone dying. This was especially the case with farmers who rented land. They worried that when a landlord died, the heirs would sell the land to the highest bidder or that one of the heirs would decide to farm it instead of renting it.

"A lot of the hard feelings come from the family farm being passed to the next generation," a corn and soybean farmer observed. "Somebody always thinks they're getting the short end of the stick." There were families in his community that had not spoken to each other in years, he said. One son would get the farm and another wouldn't. Or one heir would want to sell the farm and another wanted to keep it. "Mom finally passes away after the two boys have been farming the land for years. Which one gets the farm? It's tough."

The frequency of such conflicts was one of the reasons third-party mediation was evident among many of the farmers we interviewed. It was not the formal arbitration characteristic of labor-management negotiations or divorce settlements. But it did involve a third party whose role was to render an authoritative opinion.

More of these functions were being performed by professional experts of various kinds than had probably been the case in the past. Instead of an authoritative opinion being rendered by a parent, an older sibling, or an elderly relative, the parties involved were financial advisors, accountants, brokers, bankers, farm loan officers, lawyers, soil scientists, and agricultural economists. One of the county extension agents we talked with said her work over the past decade has increasingly focused on resolving conflicts, particularly over rental and lease agreements. She recently sponsored an event that drew more than two hundred farmers to hear a mediator discuss effective ways of communicating farm values, moving past family conflicts, and planning for the future.

In rare instances the third-party mediator was a family therapist. One of the more emotional accounts a farmer shared involved a falling out with his sister and between his son and his sister's son. A misunderstanding about a piece of land occurred that resulted in words being spoken that should not have been, according to the farmer, followed by the parties refusing to speak with one another for several years. Tortured by their fractured relationship, the farmer and his sister sought help from a family counselor. He and his sister have reconciled, but the sons are still not speaking.

A community leader we spoke with who had served as a third-party mediator thought the financial conflicts usually masked something deeper. In his experience the parties involved were often reluctant to talk about financial arrangements because they wanted to avoid these other topics. On the surface the issue might seem to be about a parent treating a grown son or daughter unfairly, for example, but underneath it had to do with someone being a jerk. "There's a history to it," he said, "and people don't want to talk about their dirt."

WOMEN'S ROLES

Although a majority of the farmers we spoke with were men, their wives and mothers were actively involved in farming as well. We spoke with a number of the women and asked them about their roles. The women played an important part in handling the routine activities of farming. There was also clear evidence of the ways in which these roles are changing.[11]

Among the older women, especially ones whose husbands were semiretired from farming, and in the stories farmers told about their mothers and grandmothers, the role of farm women was chiefly to cook, clean, manage the garden, raise the children, and do whatever else was physically possible. That ranged from milking cows to feeding pigs to driving a tractor. "The women, they were just as much a part of the farmwork as the husband was," one older woman observed. "When I met my wife," a farmer in his early sixties recalled, "she was driving a tractor with no cab and pulling a combine, and I thought, 'This is going to be a good deal because I get a wife *and* a hired man.'"

Women who currently helped with farming on larger and more mechanized farms played less of a role day-to-day in performing these traditional tasks but continued to be essential during peak seasons and for other specialized activities. Mary Jorgensen—Clay's wife—has always pitched in, for example, during harvest by helping drive a truck or the tractor pulling the grain cart. Having grown up as a "city gal," these were tasks she could handle and she enjoyed being involved during harvest. She also operates the big riding lawn mower every week from spring to fall, keeping the grass mowed on all sides of the house and around the other buildings.

The majority of younger women we spoke with who lived on farms held off-farm jobs. Arlene Jorgensen, for example, has a college degree and drives to a town of about seventeen thousand people some thirty miles away where she holds a full-time job doing the professional work in which she earned her degree. Clay Jorgensen says his daughter-in-law doesn't care much for farming. Neil Jorgensen says his wife is happier working in town than she would be on the farm. He is happier too. The extra income is crucial to the family budget, especially in lean years.[12]

The women we spoke with who had college degrees felt not only that they were contributing to the family budget but also that they were able to have a fulfilling career of their own. The income and working in town contributed to their sense of having an independent identity. Being away from the farm was sometimes an emotional lift in itself. Having gone to college, however, meant that these women more often than those who had not gone to college had grown up in a different community from where they were now living. That was sometimes a difficult adjustment. They felt like outsiders even after living where they did for several decades.

The income women earned from off-farm jobs generally went directly to the family budget. The money paid for health insurance, covered household bills, or helped with farm expenses. There were a few examples, though, of what are sometimes called "special monies." Special monies are funds set aside for particular activities that families do not consider part of their regular expenditures. In earlier generations special monies included egg money or pin money that women used for discretionary spending such as ladies' missionary society contributions or small personal items. In our interviews special monies were used for family vacations, college funds, gifts, and parties. One of the women's off-farm earnings went entirely for children's wedding funds.

Although they may hold off-farm jobs, the younger farmwomen we spoke with usually doubled as partners responsible for some of the farmwork as well. Mrs. Rayburn's role in the cotton farm she and her husband operate is fairly typical. She and Mr. Rayburn met in college. Her degree, like his, is in agricultural economics. Unlike her mother-in-law she does not raise chickens or tend hogs. She does not cook for ten or fifteen hired hands during peak seasons like her mother-in-law did. Instead, she works at a bank where she handles farm loans. Her understanding of farm loans is vital to her and her husband's own farming. After banking hours and on weekends she frequently runs farm errands as well.

Mrs. Bower is one of the few younger women we spoke with who does not have an off-farm job. Having grown up on a farm, she prefers to remain actively involved in the farm's day-to-day activities. She helps with the fieldwork during harvest and makes decisions with her husband about major purchases and farm finances. She also drives one of the tractors. Power equipment and automatic steering has made that easier. But, interestingly enough, having larger and more expensive equipment has also made her more reluctant to use it. "I'd rather drive something smaller," she says, "because it's not as scary. That's why I'm not driving as much."

Whether they held off-farm jobs or worked alongside their husbands on the farm, the women we interviewed acknowledged that farm life was challenging. The family income from farming was often meager or went into land and machinery instead of household improvements. Unless an off-farm job provided it, health insurance premiums had to be covered by the family budget. Rural television, cell phone, and electrical service was often less reliable than in town.

The challenges were not only financial. The women frequently mentioned seldom seeing their husbands as one of the most significant drawbacks of farm life. The men were away in the fields operating large equipment from early morning to late at night for weeks on end. It was also difficult to make plans because farmwork had to be done when the weather was right and nobody could predict exactly when the weather would be right. As one farmer observed, "Mother Nature is your first wife."

The family conflicts that men described were sometimes even more acute among farmwomen. There were two reasons for this. One was that farming in the areas we studied tended to be patrimonial, meaning that land was passed from fathers to sons. As a result, farm families were also patrilocal: they lived closer to male relatives and ancestors than to female relatives and ancestors. This meant that the women were farther from parents and siblings and had to adjust to living near their in-laws.[13]

It also meant that women who had been raised on farms were less likely than their brothers and husbands to have grown up thinking they wanted to spend their life living on a farm. They had not worked alongside their fathers or expected to inherit the home place. Their childhood dream had been to do something else. How their life turned out left them feeling conflicted.

Their husbands knew this. A farmer we talked to whose wife had been raised on a farm, for example, illustrated the point in mentioning that his wife worked in town at a low-paying job "just to try to keep herself busy." He explained, "She doesn't really care about the farm. It isn't her, oh, she tells me she'd sell the cattle, and I don't know. It isn't her love."

For women who truly loved farm life, there were still difficult adjustments. "Especially in a farm situation, the wife would be the one that came from a different community to make her home where her husband has been," Mr. Hebner noted. "He knows the problems that go along with that particular community and is used to the family relationships he has had. So things that have just been normal for him might be very new for his wife."

The other source of conflict for women was that more of the business arrangements were handled by men and thus benefited from semicontractual understandings, whereas the women's relationships focused more on family activities that were harder to translate into formalized arrangements. For example, women we spoke with described tensions with in-laws over styles of child rearing and discipline, babysitting, holiday customs, and cooking. One woman, for instance, described feeling belittled by her mother-in-law because of how her pies turned out. Another woman said her husband seemed to still prefer his mother's cooking.[14]

"When we milked cows, I'd get out there to help with the milking and I'd say good morning to his dad." This was another memory from the woman who recalled how her father-in-law resisted the idea of switching over to milking machines. "What's good about it?" her father-in-law would grumble. "One time I stopped over at their house and I said, 'Oh, that was a nice rain.' 'Well, it was too much rain!' He had a negative attitude about a lot of stuff."

Another woman described the tension that developed between her and her husband trying to run a dairy farm in their twenties. They were working night and day, raising children, and hardly making enough to cover expenses let alone hire anyone to help. "It just put a lot of stress on the family," she says. It was the principal reason she and her husband got divorced.

Mary Jorgensen, Clay's wife, and her daughter-in-law Arlene get along fine and are happy to be living on farms that are little more than a mile apart. The tensions that Clay and Neil have had to work though have nevertheless made problems for the women as well. When disagreements arose about how the farming decisions should be made, the women sometimes found themselves being accused, or feeling that they were being accused, of causing the trouble. It was hard to resolve these difficulties because the accusations were seldom made face to face. Arlene was glad to have as much independence as she did through her job in town.

In another instance the mother-in-law did the farm bookkeeping but refused to share any of the details with the daughter-in-law. The mother-in-law knew it would be better for both families to have greater financial transparency. But she did not feel she could trust her daughter-in-law. The reason was that the daughter-in-law shopped at an upscale department store, while the mother-in-law thought the daughter-in-law should be shopping at Walmart.

A farmer who farmed in partnership with his brother acknowledged that disagreements with his brother sometimes led to tensions between the wives and between wives and husbands that were even more acute. His wife would stand up for him, he said. She felt he was not getting a fair shake in the partnership. And then words were spoken that left hard feelings.

Although nobody made the specific connection, it was evident that farming being "in the blood," so to speak, affected the relationships between men and women. For men the farm itself and patterns of farming were literally shared or handed down among blood relatives. The women were usually present by marriage only. They were the outsiders. "It all has to do with respect," one woman explained. "If the in-laws are willing to value and respect the newcomer, then the family gets along fine." But others "regard the newcomer as an outsider," she said. In those cases the newcomers "aren't really family."

There were also instances in which the wife invested more energy in keeping the farm relationships intact than her husband, almost as if she had to because of being perceived as an outsider. "I got along with your dad better than you did," a woman in her seventies exclaimed. "That's no lie! He would be the first to admit it because after your mom died, I pretty well took care of him."

Another example was the story a farmer told about his mother, who had been the driving force despite being raised in a city and marrying into the family. His father was following the family tradition and was farming inherited land but did not have a passion for farming. "He'd rather go to town and play cards at the pool hall," the man recalled. In contrast, his mother was the one who got up every morning at six, got the children off to school, saved her pennies, and pressured her husband to get the farmwork done.

This was an instance in which the farmer telling the story held his mother in high regard. Husbands' accounts of their wives' contributions were generally similar. They acknowledged the care their wives had given to aging parents and the difficulties adjusting to in-laws and a new community. "My father lives with my wife and I," a corn farmer noted, "and it's got to be very tough for her. She's the truest form of a walking saint I'll ever know." And if their wives and mothers were not saints, the majority of men regarded the women in their lives as the bedrock of family values, emotional support, and household management in addition to the roles that many of the women played in performing on-farm tasks and earning off-farm income.

But those views were not universal. "The guys get along real well," a man in his late fifties remarked. "But the gals go through their time when they won't talk to each other. Or they'll be talking and you'll hear something like, 'How come you got a new car and I still drive this old beater?'"

"I've lived around a lot of women," another farmer explained, "and I'm telling you they will cat fight over some of the stupidest things I can think of. A man will come out and tell you what he thinks, but a woman will just stew and back stab."

CHILDREN—THEN AND NOW

Like other parents, farmers insist that they want to do right by their children. They want them to grow up safe and happy. A refrain almost all of the farmers we spoke with echoed is that farms are good places to raise children. Although people in suburbs would undoubtedly say the same thing about where they live, the arguments farmers give are usually different. Farmers emphasize the virtues of raising animals, seeing crops grow, learning to do chores, and working at farm tasks alongside parents. As farming changes, these themes are also changing.

"We chopped weeds in the fields," a cotton-belt farmer who grew up in the 1950s recalled, referring to himself and his two younger siblings. He said his mother would plead with his father, "Just let them work in the mornings while it's cool." It was not that cool even in the mornings. He still remembers yelling at his siblings when they missed weeds. And yet in retrospect his memories of these times are good. "I didn't mind doing it, really. It was okay. It beat just sitting in the house."

"I was five years old when I started driving the truck," a fourth-generation corn and soybean farmer who grew up in the 1950s recalls. The memory amuses him now. "I stood on the seat and drove on the road." The truck was in low gear. He basically steered while his father walked alongside and loaded hay bales in the back. By the time he was nine he was running the combine.

A dairy farmer whose ancestors had been farming in the area since the late 1600s recalled fondly his childhood in the 1960s. The best part was driving the tractor and learning about machinery. But he also cut wood for the boiler to heat water for the cows, carried wood, lifted heavy milk pails, and helped clear the field of rocks. He is unsure exactly how to describe those activities, other than to say that they somehow made him feel real.

Farmers' childhood memories of being raised on farms include the same warm nostalgic images evident in other people's memories. The images are often of favorite pets, friends, time to play, home-cooked meals, and being tucked into bed at night by loving parents. The memories are often vivid, reinforced by living in the same house or on the same farm as parents or grandparents did. The grandfather played the fiddle right here. The grandmother baked cherry pies there in the kitchen.

"Honest to God, the school bus driver thought I lived in the garage," a fourth-generation livestock farmer says. He laughs at the memory. For six weeks each fall the family and several hired men ate meals in the garage. These were the weeks they chopped corn for silage to feed the cattle. Eating in the garage was more efficient than cleaning up to eat in the house. The man has fond memories of those meals.

The similarities with childhood memories in urban neighborhoods are evident. The stories include friends and siblings, favorite haunts, games, and family meals. The difference is that farm memories include farm animals, chores, and living in the country. The stories are of walking through the pasture, calling the cows, weeding the cotton, and driving the tractor. The memories are distinctly about farm life. They clearly associate childhood with a particular place.

Some of the childhood memories we heard in talking with farmers also emphasized risk. Steering the truck at age five or driving the tractor at age six involved an element of risk. So did milking cows and feeding cattle. Farm life could be dangerous. Accidents happened. Animals died. Living in the country was edgier than living in town.

An unusual story about risk was given by a cotton-belt farmer who remembered sleeping outside with his brother and sister and hearing a terrible cry in the night that sounded like a woman screaming. In the morning they learned that the wicked scream was a mountain lion prowling for cattle. The same farmer recalled how bad the dirt roads were, how isolated it was to live on a farm, and how difficult it was to get into town when the roads were muddy.

"We grew up with a set of rules," a wheat-belt farmer recalled, referring to the value he sees in interaction among children, parents, and grandparents on farms. "If you broke some of those rules, it was a fatal mistake. You would be killed. That's how serious it was. When dad said, 'Don't touch that,' you didn't have to wonder if he was just kidding."

Risk was evident too in the stories of growing up milking cows, feeding animals, and driving tractors. Parents worried. Children knew there was danger involved. It was good to work hard and to take on adult tasks at an early age. But these were maturing experiences because they did involve risk. That made them seem more authentic than merely engaging in childhood play.

The measure of authenticity, as farmers describe their own childhoods, is not only that farm life seemed to fit their personality. Or that they enjoyed being around farm animals and machines. Authenticity also meant learning the practical skills associated with farming from parents and grandparents. These were often skills that launched their own careers in farming.

A woman in her forties who farms more than 3,000 acres with her husband in wheat and cattle country says her husband worked at repairing an old tractor out behind the shed as a boy. As a teenager he was able to rent a hundred acres from a neighbor. She grew up around cattle and learned early about raising calves. She also picked up skills that have helped her do the farm bookkeeping.

A corn-belt farmer talked about getting his start working with livestock by raising sheep during junior high and high school. Like many his age, he belonged to the local 4-H club. The sheep project included shearing and carding as well as feeding. One of his sheep won a blue ribbon. Although he no longer tends livestock, he feels the experience taught him some of the skills that farming involves. During his first years in farming he had nearly two hundred ewes.

"Our family was very open," a third-generation cotton grower who became a farmer to carry on the family tradition explained. As a child he listened to long conversations about business decisions, modifying used equipment to make it more efficient, and understanding new technologies. He grew up respecting his father and grandfather. Early on he decided he wanted to be just like them.

The relevant knowledge in cases like these was more than knowing how to replace the spark plugs on a tractor or treat a cow's mastitis. It was also knowing the rhythms and habits of farming, the timing involved, and having the confidence to make grown-up decisions. It was being part of a family that had a reputation in the community of being good and honest farmers.

Few topics show how farm life is changing quite as clearly as the arguments farmers give for why farms are good for children now. When farmers describe their own childhoods, the ones who grew up on farms (as most did) describe following their parents around, learning to milk cows and drive tractors, and acquiring skills and habits that proved valuable to their later success in life. Farmers' descriptions of more recent childhoods and narratives about their children and grandchildren make different arguments. In these accounts farming is still good for children but for reasons that have less to do with farming itself and more to do with simply living in the country. Cats and dogs have replaced cows and pigs.

Clay Jorgensen remembers when he was growing up that the farm was almost like a magical place where he could run free, do what he wanted, and have fun feeding the animals and learning to drive the tractor. Growing up wanting to be a farmer was as appealing as wanting to become a cowboy or a fireman or a pilot. Nowadays, though, he thinks even farm kids have more interesting things to do, such as ball games and swim meets and other activities in town. He and Mary spend a lot of their time driving the grandchildren to those activities. Mary agrees. However, she thinks farms are still good places to raise children. She says hers learned "a good work ethic" and have become "very responsible adults" as a result. Notably, it is the work ethic rather than specific skills that comes most readily to her mind.

Mr. Rayburn was one of the farmers we spoke with who saw greater continuity across generations in children's farm experiences. He thinks farm kids have always been drawn to activities in town and town kids have wanted to see what happens on farms. On balance, he says farms kids have had the better deal because they could easily go to town whereas town kids had fewer opportunities to visit farms. He reports a story of his father's to illustrate the point. When the older man was in school, the town kids were always jealous. They had only white bread and bologna for lunch. The farm kids had it better. They had homemade biscuits and sausage and gravy. Mr. Rayburn figured farm kids still had the upper hand.

His own children were now in high school. When he was growing up, he began driving the tractor when he was eight and by the time he was ten was cultivating the fields. At age thirteen he was already driving the pickup to town. "It made you feel grown up," he says. He has given his children some of the same experiences. They started driving the tractor in grade school like he did and the truck a few years later. Farm life is good for kids, he says, because they "mature a little faster."

The Rayburn children are less involved in farm life, though, than Mr. Rayburn was as a teenager. The difference, he says, is that "there's more things going on nowadays; more of a fast-paced life." Most of the farm equipment is too large and too expensive to trust to the children. The children have their own activities at school and in sports and scouts. They hardly have time to do any of the farm chores. So there are still opportunities for farm kids to do things that town kids cannot. But the biggest difference, Mr. Rayburn says, is that farm kids have more freedom. With the exception of his mother's place just down the road, the nearest neighbor is three miles away. That means a "sense of freedom" for his children. "There's a large area around here to go kick the can."

Other farm parents mention the value of learning common sense. They mean simple things like keeping a budget, balancing a checkbook, or changing a flat tire. Farm life, they say, may be isolated, but they see that as a good thing. Isolation keeps children from getting into trouble. It somehow is even more real. Not a dream world. A place to be in touch with nature.

The difference evident in all these narratives is that the ones about early days emphasize specific farm skills whereas the ones about recent times focus on generalized traits. Knowing how to milk a cow, groom a sheep, or grease a combine has been replaced by understanding the value of work, being honest, growing up authentic, and having freedom to roam.

It is not surprising that farmers still think farms are good places for children, despite the changes taking place in farming. The shifting narratives nevertheless pose questions about the future of family farming. Farmers themselves wonder if any of their children will farm. The high cost of machinery and land is not the only impediment. It is harder for children to emulate a parent they never see because that parent is planting a field twenty miles away than it is a parent plowing the back forty.

"There's tremendous stress," a farmer in his fifties whose two sons are in high school says. He feels it has become harder to involve children in farm activities than when he was growing up. Parents are more often working in fields miles away. The huge equipment is not only more expensive but also more dangerous. Kids are not allowed to run it.

Farm families are keenly aware that the skills and experiences valued in the wider world are not ones that farm children can easily attain. This realization is hardly new. For at least a century the majority of children raised on farms have grown up to pursue careers other than farming. Farm parents hoped that good high schools and college training would open the doors. Today the awareness includes concerns about children going to small-town schools having fewer opportunities for advanced placement classes and farm families having limited chances to travel, live abroad, and gain exposure to cultural diversity.

The "balancing act," as one woman described it, involves trying to provide children with some of these opportunities while hoping that a farm background instills such lasting values as self-reliance, personal responsibility, a strong work ethic, and love of the land. Another parent stressed the value of children learning to be happy with less. The big problem with society, she thought, was materialism. She figured farm children had a better chance of learning different values. She hoped so at least.

Many of the farmers we spoke with described grown children who had left the farm for good. The ones with ample land and capital were more likely to have at least one son or daughter still in farming. Others were pleased to have children in occupations that at least carried forward some of the family traditions. Working for John Deere was a way to make use of the mechanical skills learned on the farm. Teaching agronomy or agricultural economics kept farming in mind. So did living in a small town and teaching, running a business, or working at a hospital.

A farm background in these instances was at least partly replicable by virtue of working in agriculture-related occupations or living in a rural area. Although the farm population was declining, its cultural influence was being carried forward in these larger ways.

The more difficult aspect of farming to replicate, farmers explained, was the strength of character they believed to be best instilled through disciplined habits associated with physical work. They considered it beneficial in their own experience to have grown up milking cows, lifting hay bales, walking the fields, and hoeing weeds.

The closest activity they could imagine to this kind of physical work was sports, and for that reason many of them encouraged their children to be involved in school athletic programs. They worried about children who spent all their spare time watching television, playing video games, or for that matter reading.

They worried about their own children in this regard as well. If farming was becoming mechanized to the point that physical work of the kind that children could learn was no longer possible, they felt that something valuable was indeed in danger of being lost.

GROWING OLD

Implicit in then-and-now comparisons of childhood farm experiences is the fact that many of today's farmers are aging. They remember simpler times when farms were smaller and families were larger. They have witnessed dramatic changes in farming during their lifetimes. Their continuing involvement is an important aspect of family farming.

The fact that many farmers in their sixties and seventies are still farming has important implications for farm families. It sustains the memories and traditions that are so important in farming communities. It provides for continuities in family land succession. Intergenerational relationships are clearly involved in many of the partnership arrangements and in some of the conflicts that farmers describe.[15]

These arrangements resemble the assistance young adults receive in other settings from parents and grandparents. Research demonstrates that young adults in urban settings commonly benefit from parents and grandparents who provide financial assistance, co-sign on car loans and home mortgages, give career and marital advice, help when newborns arrive, and care for children in lieu of daycare and after-school programs.

An important difference in farming communities is that generational succession in many instances literally involves grandparents providing supplemental farm labor until young adults take their place. "I gradu-ally stepped in to my grandpa's role as he got older and as I got older," a wheat-belt farmer explained. "I didn't see that as clearly then," he added. "I thought I was just helping Dad, but now that I look back on it there was a one-to-one transition there."

Generational succession like this has several practical implications. The fact that it occurs gradually is important. A gradual one-to-one transition permits the younger farmer's income and investments to increase slowly as the older generation's decreases. It also increases the level of personal identification across generations and the opportunities for intergenerational transmission of knowledge.

The rewards for older farmers themselves are evident among the farmers we spoke with as well. The benefits of staying in farming past normal retire-ment age are not entirely financial, even though income supplementing So-cial Security benefits is typically important. The rewards are also emotional and social.

Among the older farmers we spoke with, comments about declining strength were not uncommon. Nor were remarks about passing from the scene, turning things over to younger farmers, and dying. Farmers were well acquainted with the idea of being "put out to pasture." They worried about sitting with idle time on their hands. They wanted to still somehow be useful. "I just hate growing old," one farmer complained. "I know a lot of worthless stuff. I know about a lot of things, but as far as doing me any good from now on, I don't know."

One of the significant benefits of family farming that older farmers men-tioned was the opportunity to still be useful. They did not feel that the knowledge they had was entirely worthless. Perhaps they did not under-stand the latest technological innovations, but not everything about farm-ing had changed. There were still things they could do and ideas they could contribute.

Men and women alike expressed this view. The ones still active in farm-ing worked with equipment that made it easier for them to be involved. The ones who were less than fully involved drove trucks during harvest, fed cattle, and ran errands. They provided meals and childcare.

It was especially rewarding to live close by, they said. From time to time at family gatherings the old stories could be told and retold. "There's a chance to tell somebody that you've done what they're getting ready to do and you know how to do it," one farmer noted. A piece of practical advice could be passed along. A neighbor could be visited. Children and grandchildren were there to carry on the family traditions. The familiarity of the land provided continuity.

A corn-belt farmer who was pushing eighty when we spoke with him was especially keen on the continuities he saw in farming. Two of his sons were

44

now farming his land. His daughter and her husband farmed a few miles away. His younger brother and a second cousin farmed in the area as well.

He knew the stories of previous generations by heart. His dog lay patiently on the floor while he talked about his family history. How his grandfather had nearly lost the shirt on his back in the 1930s. How his mother heard the coyotes yowling on the farm at night.

He felt good still being active even though he was semiretired. He said he had never been very good at repairing machinery but did a decent job of managing the books. He used the laptop sitting open on his desk to check the daily markets. He and his wife were the go-fers whenever a farm errand had to be run. It pleased him that one of the grandsons is majoring in agronomy and is talking about farming when he graduates from college.

In all these ways, family relationships were vital. Farming was a business, to be sure, but it was more than that. It was integrally connected with the relationships in nuclear and extended families. The relationships thoroughly shaped its meanings. They were the vessels in which traditions were stored, the sources of occasional conflicts and misunderstandings, and the reasons farmers hoped their way of life could be passed on to their children.

NEIGHBORS

2

Everybody knows everybody else's pickup around here. We help out, but not like we used to. Those days of a community coming together unfortunately are past.

—Cotton-belt farmer, male, age 64

If you watch a group of farmers talking, it's like a grand poker game. Everybody's trying to say what they say without really saying it.

—Truck farmer, female, age 53

Farming takes place in communities. This is a point that would have gone without saying a century ago. Now it is a reality that is easier to miss. Passing farms as one zooms along an interstate highway, a casual observer would notice a farm here or there and perhaps not see another one for several miles. It might seem that farmers had few neighbors at all. A more knowledgeable person might peruse government statistics about the number, size, or average income of farms and conclude that each one is a stand-alone business unit. Again, the farming community could easily be missed.

I wondered what farmers would say when asked about their neighbors. The academic literature has been interested in Americans' declining sense of community. Writers on one side of the debate argue that people think community is declining because it really is. The other side says no, the perception is rooted only in a nostalgic view of the past.[1]

The same arguments can be made about farming communities. Perhaps community here is declining, too, as fewer people live on farms. Or perhaps neighborly connections have remained vibrant, strengthened by extended family relationships and by the fact that farmers and their neighbors do similar work and benefit by sharing ideas and information.

The farmers we spoke with said one of the joys of farming is having neighbors they know and interact with—neighbors doing similar kinds of work and sharing similar values. For many farmers, some of the neighbors are blood relatives. That affects the dynamics involved. Farmers are also unusual in that many of them have lived in the same place all their lives and expect to stay there until they die. That also shapes how they perceive and value their neighbors.[2]

Communities have histories, just as families do. Living in a farming community for an extended period intertwines personal and family biographies with community history. Farmers characteristically know when the first settlers came to their area, when the nearest town was laid out, and who the descendants are who still live in the area. Part of what makes the community home is its familiarity.

Familiarity is inscribed in visual reminders and in narratives about the connections among people and places. The land is special, the story goes, because the first settlers came from Europe, where it was impossible to purchase land. The old barn over there is where a profitable dairy herd used to be milked. The farmers who lived there were good neighbors. That abandoned farmstead is where a discouraged neighbor committed suicide. Being part of the community means being interested in these local details and being able to pass along the stories to newcomers and the coming generation.[3]

When neighborly relations exist for a lifetime, the social norms that govern these relationships are an important aspect of community life. Farmers are keenly aware of these norms. They know when it would be unthinkable not to stop whatever they are doing to go help a neighbor. And yet they also know how to keep these expectations about helping under control so that their own work actually gets done. They know when to chat with neighbors over coffee and when not to chat or when to avoid certain topics. As one farmer observed, "you leave your pickup running" if you do stop to chat with a neighbor. That means you are not intending to talk for long.

Neighborly relations are being affected by the changes taking place in farming communities. As technological information becomes more specialized, norms are changing about how neighborly conversations fit into the sharing of such information. As the cost of machinery escalates, expectations about the sharing of equipment are shifting as well. The same is true of changing relationships to the land itself. As land prices increase and as land becomes harder to purchase or rent, one's neighbors become one's competitors. Some of the farmers we spoke with were surprisingly candid about these competitive relationships.

One other change that has been affecting farming communities is new ethnic diversity bred from recent immigration. Census statistics show that even small rural communities are increasingly the location to which new immigrants are drawn. The attractions range from seasonal work on farms to steady work at large feedlots and meat processing facilities to inexpensive housing. The statistics, though, do not illuminate whether neighborly relations in farming communities are changing as a result. One possibility is that greater opportunities for interaction do in fact affect relationships and attitudes. An equally plausible scenario, however, is that proximity does not result in interaction.

Farming communities offer a revealing glimpse of expectations about neighborliness that are in many ways unchanged from previous generations. At the same time the story is not primarily about the strength of informal ties among friends who just happen to live in the same vicinity and share common interests. It is just as much about formal relationships that are defined, cemented, and regulated by formal organizations.

EXPECTATIONS ABOUT SHARING AND HELPING

Nearly all of the farmers we spoke with asserted that neighbors in their communities help one another. Indeed, one of the best things about farming, they said, was having good neighbors. Being a good neighbor meant giving help when help was needed. That was the norm. Their best neighbors, they said, customarily lived up to that norm. They had stories to prove it.

"If somebody dies or gets sick during harvest time," Mr. Rayburn observed about his fellow cotton farmers, "we'll all load up our equipment and go help." During the years he has been farming that seems to have happened at least once a year. The neighbors will go strip the person's cotton and take it to the gin. Just recently a farmer in the community was hospitalized for several weeks following a heart attack. The neighboring farmers planted his cotton for him. "It's customary," Mr. Rayburn explained. "You take your equipment and you go take care of that family when something happens."

A couple in their late sixties who were still harvesting a thousand acres of wheat each summer told us this was the kind of community they lived in, too. The husband developed a staph infection late last spring and was hospitalized for eight days. As harvest approached, he was too weak to get from the house to the shop, let alone operate a piece of machinery. The neighbors came over and did the work. "It really made us realize how blessed we are," the wife recalled.

"When someone comes down with cancer or loses a wife or something tragic all of a sudden happens and its harvest time," a cotton farmer in his early forties asserted, "you will see eight or ten machines in one field. It may be only a hundred-acre field, but you'll see that many machines in that one field."

An older corn-belt farmer we talked with described an event that stuck in his memory as an example of neighborliness even though it happened years ago. As a young farmer, he was teaching school to help supplement the meager earnings from the farm. One evening he came home from school and saw two combines and three trucks in his field. The field was planted in pinto beans, which were particularly susceptible to wind, and the wind had been blowing hard that day. He wondered what was going on. He had only one combine and one truck.

When he got to the field to investigate, his neighbor greeted him. "The wind was blowing so hard, those beans were trying to roll into town on their own," the neighbor said. "We've been chasing your beans for you all afternoon. Some of them got away from us, but we saved most of them."

The point of the story was not to suggest that all neighbors would have done the same thing. It was rather to show that some neighbors violated customary norms of neighborliness by going beyond the call of duty. "He was a very good neighbor," the farmer said. "He did a lot of things that an ordinary neighbor would not do."

Neighborly assistance exceeding ordinary norms was the kind that incurred special debts. The recipient was expected somehow to reciprocate. The same farmer told another story to illustrate the point. "I finished combining one fall," he continued. "I cleaned up my truck and put it away in the Quonset. My neighbor came over one night and asked, 'Where's your truck?' 'In the shed,' I said. 'I put it there some time ago.' 'Well, let's go take a look,' he said. We went out and opened the big door and the truck was gone. 'I've been using it for a week to haul wheat,' he said." They both laughed about it in following years. That was the kind of trust that cemented their relationship.

A wheat-belt farmer in his sixties described a similar relationship with one of his neighbors. The neighbor had recently died. The farmer who had been his neighbor mourned the loss. Their relationship had been special— "tight," to be exact. They knew each other well enough to exchange advice and information on a regular basis. The farmer said that was unusual. For the most part, he said, farmers did not talk much about their business because they knew how easily gossip spread. "They're quiet that way," he explained.

Other farmers made it clear that sharing and helping followed particular social norms that were understood in the community. A neighboring farmer who deserves help is one who has been seriously injured in a farm accident, fallen ill at a bad time of year, died, or for some other reason is facing serious problems that cannot be avoided. One of the farmers we spoke with expressed the idea succinctly. When asked if farmers in his community help each other out, he hesitated for just a moment and then replied, "If need be." His wife elaborated, "If a farmer's had bad health and can't harvest a crop, the other farmers will get out there and harvest his crop for him."

A farmer who lives near a town of about a hundred people that caters to the small crop and grassland farmers in the area told of an incident that in his view clearly illustrated the social norms governing neighborly assistance. He said farmers in his community helped one another when something

49

tragic happened, but he thought assistance of this kind was also tempered by a strong and perhaps unhealthy dose of materialistic self-interest.

The incident involved the deaths of a farm couple in their mid-thirties in a tragic automobile accident. Immediately the neighboring farmers pitched in. They fed the cattle and harvested the fall crop. Before long most of the assistance stopped. It seemed, he said, that the assistance was being given by neighboring farmers in hopes that they could purchase the land.

The farmer telling the story acknowledged that perhaps he was being cynical. Still, he thought neighbors were not as willing as they used to be to help out when a tornado struck or a barn burned down. Now it was like, "Gee, the neighbor had tough luck. I hope he goes broke and then I can buy his land." In short, good neighborliness meant showing the milk of human kindness, but all the while being aware that neighbors were self-interested as well.

Other farmers were less jaded but agreed that expectations about sharing and helping included strict limits on how much was enough. The prevailing ethos emphasized hard work and personal responsibility. Farming was a business. Labor and machinery were valued inputs that were to be freely shared only within limits or on an emergency basis.

These expectations vary somewhat, we noticed, in different kinds of farming communities. They varied especially in terms of whether the definition of neighbors included townspeople or only neighboring farmers. Farmers producing wheat or corn may have had little need to interact with neighboring farmers or with townspeople, whereas truck farmers who sold directly to the public usually had business reasons for cultivating relationships with friends and neighbors.

One of the truck farmers we interviewed lamented the fact that, in his view, there was a growing gap between his rural neighbors and his urban neighbors, by which he meant that consumers of the fruits and vegetables he produces seem to have less understanding and appreciation of what he does than was true in the past. He works to close that gap. He sits on boards with other truck farmers in the area who discuss better ways of engaging in direct marketing.

Lately he has been hiring students from the local high school to help on weekends and during the summers. That builds relationships with families in town. It helps, too, just to spend more time in town. He says he almost hates to go to the post office because he runs into people who want to talk for fifteen minutes, but he figures it matters to build these relationships.

The important aspect of these neighborly relations that can almost be missed involves their role in community and family stories. The social ties involved are not only the kind of links that a statistician would map in describing a social network. They rise above that, becoming the stuff of which ideas about community spirit are made. Assistance given is grist for coffee-shop tales affirming that this is a good place to live. Assistance not given when it should have been lingers as a cautionary reminder. It may be especially hard to forgive a neighbor who did not behave like a Good Samaritan when the opportunity arose.[4]

"There was a story that went along with the people in the big house over there," a farmer says as he describes the neighbors in his community. He is no longer sure of the details, but it had something to do with the neighbors pitching in when someone in that family was ill. He was thinking about it just the other day. He knows that the elderly woman who drives the white pickup to get her hair fixed is related to those people. Seeing her pickup reminds him of the stories he has heard about how the community helped when help was needed.

If neighborly assistance is especially evident at times of acute need, some of it happens in routine ways that are nevertheless significant. Hiring high school students, as the one truck farmer does, is an example. The students learn aspects of farming and come to have an appreciation for it. Many of the farmers we interviewed mentioned having learned in these ways from neighbors. For them neighborliness consisted not only of emergency assistance but also of mentoring.

An example that underscored this point came from a dairy farmer who grew up on a farm and earned a college degree in agriculture. College was "just the academic part of it," he said. "The actual experience of working with farmers really made the difference," he explained. That experience began while he was still in school and continued during his first decade in farming. He not only learned about working with animals but also acquired ideas about which farmers made money and which ones did not. In his case the neighbors were role models. What mattered was getting to know them and learning how they made decisions.

While the norms defining what it means to be a good neighbor function in specific ways, then, they demonstrate that neighborliness is indeed an important part of farming communities. Having good neighbors who help during emergencies is one of the ways that farming communities protect themselves against these events. It matters if at least one or two neighbors are especially trusted and helpful. Young farmers gain experience and insights by working for neighbors. At the same time the norms limit the amount of neighborly assistance that good neighbors are expected to give. Farmers know that neighbors are competitors and that being self-sufficient most of the time is important.

ADAPTING TO NEW REALITIES

The realities to which farm families are adapting are that many farmers feel they are too busy to spend time with neighbors and are involved in such highly technical farming decisions that neighborly advice may not be

51

trustworthy. Neighborly relations are especially affected by the fact that fewer people live in farming communities.

"It was nothing for a neighbor of any kind to come over and stop and visit for a few minutes." This is Neil Jorgensen describing farm life when he was growing up. In those days the farmers in his community also shared work when some extra hands were needed to herd cattle or put up a new grain bin. "Just let me know," they would say, "and I'll come over and help you." Mr. Jorgensen still tries to be that kind of neighbor. "If a neighbor needs help, I'll go help them, and I don't expect anything from them. A 'thank you' is good enough." But he says that does not happen very often and there are only a few neighbors who seem to feel that way. "It's nothing like it used to be," he observes. Everybody is too busy with their own work.

"I remember the old days," another corn and soybean farmer says. The old days in his case were the 1970s. "You'd be working the field and neighbors would stop the tractors, walk over, and visit." Nowadays that does not happen. "You don't do that anymore." About the only time he talks with neighbors is at the machinery dealership when he's there for parts. "Those relationships just aren't there quite as much as they used to be."

In the part of the cotton belt where the Rayburns farm, a typical farmer usually farms at least a thousand acres. That means fewer immediate neighbors. As he says, his closest neighbor other than his mother's farm is three miles away. It also means that some of his land is fifteen to twenty miles away. He farms about a thousand acres east of town and a couple thousand more south and north of town.

He hardly knows the farmers on those sides of town. He sees them only in passing. He recognizes their trucks and waves when they pass on the highway, but that is the extent of it. He has a bit more contact with his immediate neighbors. He sees them at the cotton gin or at the seed store. "I'll run into them once every two or three weeks for sure," he says. Asked if he and his wife get together with these neighbors for social activities, he says, "Not so much."

As a mark of neighborliness, it makes sense that less time spent socializing suggests weaker neighborly relations. But socializing is only part of the story. Community is as much about knowing people and trusting them as it is about spending time socializing. It rests on seeing familiar faces, knowing who they are from having talked to them at some point in the past, or knowing that someone you know could tell you about them if it became important.

Neighborliness of that kind means that a friendly wave to another farmer on a country road is enough. Anything more would be an unwelcome intrusion on a busy day. Familiarity implies that one person would help another if there were a true emergency. Infrequent though the interaction may be, it includes a sense of togetherness. "You know what's going on with your neighbors," another farmer muses. "That's just the way it goes. I mean, if anything's needed, we help out."

It is this sense of being in it together that farmers we spoke with said they appreciate about their community and feel is generally still present, and yet seems to be less evident than it used to be. Mr. Hebner, the wheat-belt farmer whose in-laws help during harvest, gave an interesting example. He said the farmers in his community simply are less visibly present than they were when he started farming. He thinks about this when he cuts wheat at night. There used to be a kind of togetherness from seeing a neighboring farmer's combine lights in an adjacent field. Now those lights are much farther away, if they are still visible at all.

Besides having fewer close neighbors, farmers with large farms that involve partnerships and family corporations say they sometimes have quit sharing labor and machinery with neighbors because of legal and financial considerations. They may rent machinery to a neighbor or hire a neighbor to cut wheat or mow hay, just as previous generations did. They think this is happening less often, though, and when it does, formal business agreements are involved. It is almost as if farming on a larger scale necessitates a switch from informal friendly relationships to formal contractual relationships.

A county agent in the wheat belt described one way in which the trend away from farmers working together was being reversed. No-till crop farmers in recent years were starting to partner with livestock farmers who wanted to expand but did not have sufficient grazing land. The crop farmers let the livestock farmers use some of their land for grazing if the livestock farmers provided the seed for ground coverage and did the planting. These arrangements nevertheless were rare, the county agent said, because they took time, required trust, and necessitated contractual agreements.

The farmers we spoke with who still worked together or who actually shared machinery usually had smaller-than-average farms. They shared out of necessity. They could not afford a specialized piece of equipment, such as a self-propelled sprayer or bulldozer, so they borrowed or rented or co-owned with a neighbor. Or they traded labor because they farmed alone and had no hired help.

One example was the Loeschers. Although they had lived in the same community all their lives, they were the last of their extended families to have remained in farming. The dairy herd kept them busy enough from morning to night that they did little more than wave at some of their neighbors. They had good relations with one of the neighbors, though. They shared equipment with the neighbor, and the neighbor sometimes borrowed theirs.

Neighborly relations in farming communities are being affected as well by a fact less obvious than being busy. The Loeschers' experience illustrates this change. Not having extended family members in farming means that most of their relatives live elsewhere and work in different occupations. If one were to map their social networks, their most enduring relationships would reflect the geographic dispersion of their kin.[5]

As dairy farmers the Loeschers are tied to their locale because of the daily chores of feeding and milking their cows. That is less true of corn, cotton, soybean, and wheat farmers. The seasonal nature of their work allows them to spend days and weeks traveling and visiting relatives and friends in other places. A map of their social networks would reflect even greater geographic dispersion.

We asked the farmers we spoke with how much they traveled and why they traveled. A few were like the Loeschers. They seldom got away. But most said they were able to take vacations during the off-season or make day trips to visit friends and relatives. Among farmers with grown children, nearly all made trips to visit children who no longer farmed and no longer lived nearby.

A few of the larger farmers also said it was increasingly common among their peers to travel on business. They traveled to the state capital or to Washington, DC, to attend farm association meetings and were increasingly involved in international networks as well. A wheat-belt farmer whose income was mostly from cattle was an example. He owned thirty farms that included more than ten thousand acres near his home. His purebred livestock had also become popular in Latin America. He did a significant amount of export work to Brazil and Argentina, traveled there regularly, and participated as a judge at international livestock shows. For him the definition of neighbor was clearly less local than it had been for farmers in his area in previous generations.

Geographically dispersed networks are not necessarily inimical to strong local ties, but they do shift the balance of social relationships in subtle ways. While most of the farmers we spoke with had lived in one place for a long time and were intensely loyal to their local communities, they were by no means isolated. Friends and relatives and business interests linked them to a more cosmopolitan stream of ideas and influences.

SHARKS IN THE WATER

For people who live in cities and suburbs, the potentially troubling relationships that arise with neighbors usually have to do with minor violations of privacy. A neighbor plays loud music late at night, makes noise too early or too late mowing the lawn, wants to chat for an hour when you are busy, and so on. Neighbors can also be a nuisance if they fail to keep their lawn mowed or have a dog that barks all hours of the day and night. These are the kinds of disturbances that zoning laws, the police, and informal norms of common decency are supposed to prevent. For the most part people expect to have good relations with their neighbors either by being friendly or by keeping to themselves.

If farming communities benefit from neighbors doing similar kinds of work, living in the same area for extended periods, and knowing when to help and when to leave one another alone, they have the added fact—which can be a danger—of neighbors essentially being in competition with one another. The situation is a bit like members of rival athletic teams living in the same neighborhood. Except that athletic rivalries are all in fun. A better analogy might be rival storeowners living next door to each other.[6]

Farmers' competition with one another is about land. Nearly every farmer we spoke with wanted more land. When a neighbor succeeded in purchasing or renting an additional piece of ground, that was an opportunity lost for that person's neighbors. They knew it was. The opportunity might be lost for a lifetime. It had real consequences.

It also has symbolic consequences. Winning or losing in securing land was an indication of success or failure. When getting larger was the name of the game, not being able to expand meant falling behind. Or at least it felt that way. The neighbor with more land would have an advantage next time as well.

Neil Jorgensen's explanation for farmers in his community being less willing to help one another than in the past was that they are not only too busy but also more likely to view their neighbors as competitors. "Neighbors just got away from [helping]," he says. "I don't know why. I think because it's a competition thing. Everyone wants your ground that you farm now. They just can't wait for something to slip up so they can get in there and get your ground now."

Mr. Bower, the wheat-belt farmer, put the idea succinctly when asked what he saw as the greatest threat to the family farm nowadays. "The biggest threat to the family farm," he replied, "is the neighbor's family farm." He and Mrs. Bower knew this from personal experience. She recalled her dad drinking coffee two or three times a week with several of the neighbors. They sometimes rented pasture to each other and shared work during harvest. But she and her husband have not followed that pattern. They view the neighbors more as competitors.

One of the Bowers' neighbors shares these worries. She and her husband farm a relatively small number of acres and are trying to divide things to help their son get started farming. "Just about the time you think you're getting average size," the woman says, "you end up being small again. Some of these farmers are hogging up the land. The price they're willing to pay for it is just crazy! They'll steal something right out from under you. They're so greedy about getting more land."

Another of the Bowers' neighbors was in his sixties and trying to acquire additional land to bring his daughter and son-in-law into the business. He felt neighborly relations were good when long-time farmers retired and offered their land to someone they knew and respected. His best relations were with friends he had known since grade school. But he thought that pattern was less evident among younger farmers who were trying to get established. "The younger ones," he mused, "they don't want their neighbor. They want their neighbor's land." Farmers his age, he felt, were "to the place where who gives a darn." But the younger ones were "still hungry and growing."

As one of the larger farmers in his community, Mr. Lancaster was sensitive to the fact that cutthroat competition was inimical to good neighborly relations. He currently farms four thousand acres and says he could handle twice that much. But he has passed on opportunities that would have damaged relations with his neighbors. "Any time a piece of ground changes hands," he observes, "somebody's feelings get hurt." He would rather have good neighbors, he says, than another piece of ground. He knows he is in a better position to feel this way because he already has extensive land.

Perceptions about the extent and nature of neighborly competition inevitably vary depending on farmers' own position in securing land. Some of the farmers with large amounts of land were like Mr. Bower and Mr. Lancaster. They saw their neighbors as competitors. But they also thought the competition was basically fair. Large farmers won because they had more to offer, such as bigger machinery and more efficient methods. Smaller farmers saw it differently. They were more likely to decry the competition.

Perceptions also varied based on whether the competition was coming from farmers or from outsiders. Competition from neighboring farmers struck raw nerves because it implied that here was someone of basically equal status who was doing better. At least this kind of competition seemed fair, and farmers chatting over coffee and sharing similar problems took off the edge. Competition with outsiders was harder to stomach. Outsiders held an unfair advantage and were threatening the basic rules of the game.

Ralph Engstrom is a fifth-generation farmer who farms in partnership with his brother. Their primary crops are corn and soybeans. Together they farm about a thousand acres. That puts them well below average in their community. Although Mr. Engstrom recently celebrated his sixtieth birthday, he would like to expand. His wife's job in town is still essential for covering the monthly bills, and they wish they could get one of their children started farming. He and his brother would like to farm more, but land to rent or to buy is scarce. Their biggest competitor is a rich widow who lives in town and does not farm but likes to buy land.

"There's a woman out there who's got money," Mr. Engstrom explains. "If she knows something is for sale, she'll try to get it." Mr. Engstrom warms to the topic. "Her father bought into a little company about thirty years ago. Or something like that. A long time ago. It turned out alright," he says, deliberately understating the claim. "It was . . ."—and he names a well-known company now worth hundreds of billions of dollars. The woman inherited the stock when her father died. "She has a big fat operation over there," Mr. Engstrom says, "the highest priced house in the county, biggest house you ever saw." The garage is larger than most people's house. So she can buy land whenever she wants it, and she does want it. "She can be a bit rammy," he says, meaning that she strikes him as pushy or aggressive. "But when you got money, it can happen that way."

How do farmers cope with this kind of competition? Other than bemoaning it or privately deriding the competition, they handle it by cultivating close relationships with some of their neighbors. They shore up ties within the community, even though some of those ties are with competitors, as a way to gain an advantage against people outside the community. Outsiders may have more money but lack the inside track that farmers themselves have when land becomes available. An inside track works by preventing information from spreading beyond those in the network.

"You've got to watch what you're doing out there," Mr. Engstrom says. "Keep it quiet, because if you want to buy or whatever, she's got money and likes having things her way."

Cultivating a quiet relationship with one of his neighbors was how Mr. Engstrom managed to buy one of his most valuable parcels of land. The owner was "kind of a shyster," Mr. Engstrom recalls. "Nobody thought I was going to get it bought. They didn't think he would come through with the deal. But the thing is, I knew the old fellow stayed glued to the sales," he says, referring to the weekly livestock sales at the auction barn. "He'd be at the sales and you had to kind of bullshit him a bit. If you listened to him tell his stories, when he got down to selling the farm, he wanted me to have it."

Another corn-belt farmer provided an illustration of neighborly relationships being helpful in securing land. As a young farmer just starting out, he worked for one of the neighboring farmers. These neighbors were good friends of his parents and did not have children of their own who wanted to farm. So, having worked for them for four or five years, the young man was the obvious person to become their renter.

About the same time, another neighbor who was a bachelor retired and rented the young man his farm. "It all comes down to relationships," the man observed. He thought that was still true in his neighborhood. "Some people do not want to put their land up for sale or for bids. They've got maybe one or two people in mind that they would like to rent to, and you've got to build a relationship with that person." That was how neighborly relationships worked to someone's benefit.

He feared all that was changing, though. One reason was that a few of the farmers in his area had accumulated so much land that they could secure more even though they did not have these kinds of personal relationships with neighbors. For instance, he knew of a farmer some distance away who farmed between ten and twenty times the average number of acres of most farmers in the area.

"He'll just keep taking whatever he can get," the farmer explained. "He's happy with a very small margin per acre because he's farming so many acres. He'll come in and just blow everybody out on rent prices." The other change, the man said, was that more nonfarm investors were buying land. They were bankers and professionals from cities who saw land as a better investment than stocks and bonds. He thought there were tax angles to the deals as well.

As one of the principal owners of Granger Farms, Tom Granger was sensitive to the fact that smaller farmers in the community felt threatened. He acknowledged that farmers sometimes "fought tooth and nail and bid against each other" whenever land became available to purchase and rent. However, he felt that was happening less than it had a few decades ago. Things had stabilized. The few farmers who were left had been in the area for a long time. A division of labor had emerged with farmers specializing in different truck crops or operating dairies. "We're all trying to survive and be as efficient as we can," he said. "Instead of each of us knocking our heads against each other, we're working together to try to survive. I think that's a big change."

Whether it was or was not true that neighbors were fighting tooth and nail to get land, competition of this kind was clearly a development that farmers considered detrimental. It threatened the fragile expectation that neighbors were in it together and would help each other when times were tough.

In one community the farmers actually fought back. They at least tried to combat what they saw as an unwelcome challenge to prevailing norms. The problem was a family in the community that was aggressively buying and renting land. Word got around that they were stepping on toes, disrespecting the neighboring farmers, and doing everything short of violating the law to acquire land.

"Everybody says it's gonna come back to haunt them," one of the farmers in the community said. "People are talking about it negatively. Little things like that. It could be pretty big. It could affect their business and their reputation." He felt this kind of talk and the negative repercussions that could follow were still one of the ways that a farming community could protect itself from overly aggressive behavior. It was an uphill battle, he said, but a struggle that could be won as long as neighboring farmers stuck together to resist these unwanted outside influences.

NEW IMMIGRANTS

Between 1990 and 2010 more than 25 million legal immigrants came to the United States and as many as 11 million more were estimated to have come as undocumented workers. Although the majority of new immigrants settled in large metropolitan areas, by 2010 more than 80 percent of small rural towns with populations of at least five thousand included some immigrant families as well.[7]

In earlier decades immigrants who lived in small towns usually worked on farms as low-wage laborers doing such tasks as irrigating fields, feeding cattle, and helping with planting and harvesting. In recent decades large meat and poultry processing plants have become the more prominent sources of employment for new immigrants. A study of employment patterns in Kansas, Nebraska, Iowa, and several other Middle West states showed that annual wages in meat processing plants were significantly higher than among farm laborers in terms of both hourly rates and number of days per year.[8]

Recent immigration raises the possibility that farming communities are becoming more diverse racially and ethnically and, if so, that the sense of neighborliness in these communities is changing as a result. At the same time employment at meat processing plants rather than on farms may be an important dynamic in these relationships, promoting awareness of newcomers, for example, without facilitating direct contact.[9]

Hardly any of the farmers we spoke with said they knew any recent immigrants or that immigration was much of a factor in their local environs. There appeared to be three principal reasons for this lack of immigrant presence and contact. Land itself was expensive enough that immigrants were unable to purchase land, and as far as renting was concerned, landowners rented to established farmers in the area who also owned land and who had ample stocks of experience and machinery as well as good credit with local banks. Those considerations ruled out renting to recent immigrants.

A second reason pertained to farmers hiring immigrants as wage laborers. Many of the farmers we spoke with said it was difficult to find hired help and some complained about native-born people in the area no longer being willing to do physical labor or work at low-paying jobs. But instead of hiring immigrants, farmers did the work themselves and relied more on mechanized equipment. They could have used immigrants in some cases to operate the equipment but considered the equipment so expensive and requiring so much skill to operate that immigrant labor was not desirable.

That left as the only possibility for interacting with immigrants chance encounters in town, such as at the grain elevator or in church. Those encounters, though, appeared to be nonexistent. In many of the farm towns, the number of immigrants was small. Farmers talked to people they knew and figured immigrants probably did not speak English or would have been too busy to talk.[10]

The exceptions to these patterns were among dairy and truck farmers. A farmer in one community we visited said there was a dairy not far from where he lived that milked approximately 700 cows and had 1,500 more at another location. "They've got maybe twenty Mexicans working there." He said there had been no problems. "You know, they've kept their noses clean. You know, they just work." A dairy farmer in another community agreed. The Mexicans and Guatemalans in the area wanted to work and worked hard. She valued the diversity they added to the community, although she said they were not yet a "real part" of the community.

A community in which the farms specialized in apples and sweet cherries was an example of one that depended heavily on migrant workers. It had a large food processing facility that employed nearly a thousand laborers who were mostly recent immigrants as well. "When I was in high school, the demographic was 90 percent white and 10 percent black," a farmer in his fifties recalled. Currently it was 75 percent Hispanic, 15 percent Anglo, and 10 percent Asian American and African American. "You can hardly believe it," he said. "It's just amazing."

Mr. Granger's truck farm hired fewer immigrants than some of the neighboring farms in his community. The reason was that Granger Farms specialized in vegetables that could be machine harvested, whereas some of the other farms hired large numbers of immigrants to do hand harvesting.

Nevertheless, Mr. Granger considered immigrants essential. Several of his "middle management" employees were immigrants. They had been with him for fifteen or twenty years. "They can run any piece of equipment and do whatever needs to be done and take care of it," he said. They were, in his words, "absolutely vital," "irreplaceable."

Several of the farmers in Mr. Granger's community had made trips to Washington, DC, to lobby for immigration reform. The community was heavily Republican, but the view there was that Republican officials were not doing enough to ease immigrants' transition to full citizenship. As one farmer put it, "The immigration policy in this country is killing the country. We're fighting it all the time. It's very discouraging."

In farming communities with little or no direct contact with immigrants, the farmers we spoke with also held strong opinions about the potential effects of immigration on their communities and the United States more generally. The opinions were usually positive or hedged with some ambivalence.

One of Mr. Rayburn's cotton-belt neighbors illustrated the ambivalence many of the farmers we spoke with held toward new immigrants. On the one hand, he considered immigration a problem and wished the government had better policies to halt illegal immigration. On the other hand, he felt the bigger problem was that work needed to be done and native-born Americans were unwilling to do it. He respected immigrants who wanted work and were willing to work hard at these jobs. "People are struggling for food," he said. "Their families are hungry. They may not be citizens, but they're willing to do the job."

Another cotton-belt farmer expressed a similar idea. "I could care less about your ethnicity," he replied when asked how he felt about Latino immigrants in his community. "The good Lord made you and the good Lord made me. That's all that matters." One of his neighbors had just hired three men from Mexico. They were good workers. "I've never seen people work like that," the farmer exclaimed. "They worked when it was cold and wet and muddy. If I would try to get people here to do that, nobody would do it."

There were examples, though, of farmers with deep misgivings toward new immigrants. Farmers whose incomes were low and who were feeling squeezed by unfavorable markets expressed these opinions most often. Some of the farmers, like Mr. Rayburn's neighbor, felt grudgingly that immigrants were coming into their communities because American citizens were no longer willing to work hard. They acknowledged that immigrants were willing to do the work, but still thought immigrants mostly caused trouble.

"It boils down to the American people have gotten lazy," one corn-belt farmer observed. "They don't want to do physical work and they can hire these Mexicans cheaper and the Mexicans want to work. It's really affecting our communities. They bring in drugs, gun fights, knife fights. It's not good."

The other source of negative comments about immigrants reflected a kind of "us versus them" attitude, where "us" referred to the immediate farming community with no immigrants and relatively good neighborly relationships, while "them" referred to a larger community some distance away that attracted immigrants and was having problems as a result.

The farmer who said immigrants kept their noses clean at the dairy farm in his community illustrated an "us and them" view in describing a different situation some fifty miles away where a town of 25,000 included a packing plant that employed immigrants. "It has lots of problems. It seems like the Mexicans, well, they're in their own world. Sometimes in the summertime they might shoot each other and, you know, it's not good some of the things they do."

While most of the discussion about immigration in farming communities referred to immigrants born in other countries, we also encountered some concerns about another kind of new residents. These were newcomers who built or purchased houses in rural areas but did not farm. They lived on farmsteads purchased from retired farmers or on small plots adjacent to farmland. Cheap housing or the opportunity to have more space for a garden or to keep animals drew them.

Farmers' comments about these newcomers registered ambivalence. On the one hand, any newcomers helped the local store owners stay in business and contributed to the tax base and perhaps to the school population. On the other hand, farmers noted that the newcomers were quick to complain about dust, chemicals, and barnyard smells.

The newcomers contributed diversity to farming communities that was sometimes puzzling as well. "The neighbor to the west, he's not even a farmer," a corn-belt farmer explained. "The next neighbor west of him, he's a construction guy. And then the guy to the east of me, he's retired."

Having neighbors who did not farm meant having less in common. They were still the kind of acquaintances a person spoke to at the post office. They might keep an eye out when someone was gone. They were not the kind of neighbors a farmer exchanged equipment with or helped farm or had known from birth. "Not like the old days," the corn-belt farmer added.

THE ROLE OF FORMAL ORGANIZATIONS

Although the farming communities we studied still included a reasonable number of people who got together for family gatherings on holidays or whose old-timers met for coffee on a regular basis, the glue that held these communities together consisted importantly of formal organizations. Nearly every farmer we spoke with was involved in one or more of these organizations. That included even the farmers who said they were introverts and preferred to be alone.

The most commonly mentioned organizations were directly related to agriculture. These included local and regional associations of wheat growers, corn growers, and cotton growers. Many of the farmers were currently serving or had served in the past on elected boards of farmers' co-ops or the Farm Bureau. Some of the co-ops and Farm Bureau chapters had committees and subcommittees that reflected their size and federated structure. There were township committees, county committees, regional committees, and state committees. In addition, there were subcommittees dealing with conservation, water districting, animal husbandry, and specific crops. One of the farmers we spoke with currently served on eleven such committees.

Farmers' co-ops were popular with nearly all the farmers we interviewed. Among the cotton growers, co-ops usually were the organizations that operated the local cotton gins. Among wheat and corn and soybean farmers, cooperative grain elevators were the places to haul grain. Most of the co-ops also functioned as consumer cooperatives. They were the suppliers from which farmers purchased seed, fertilizer, and fuel. Some of the co-ops sold tires, batteries, and equipment for handling livestock.

The co-ops appealed to farmers because they were owned by farmers and distributed profits to members in the form of annual dividends. Member-elected boards composed of local farmers governed them. They often provided goods and supplies at reasonable prices by virtue of buying in bulk at wholesale prices. In some instances the co-ops were affiliated with statewide and national chains that owned oil refineries and fertilizer plants.

Farmers' loyalty to the co-ops went beyond the economic incentives involved. The co-ops symbolized local communities, often quite visibly through the large grain elevators that marked towns' existence. It mattered to the farmers we spoke with that the co-op had been around for a long time. They trusted it. They knew the people who worked there. The co-op was the one place in town that catered especially to farmers. If they could count on seeing any of their neighbors when they went to town, it was at the co-op.

Although they were popular, farmers' co-ops were nevertheless facing challenges in several of the areas we studied. We encountered instances of co-ops losing money, contracting, or going out of business entirely. In the Loeschers' community thirty dairy farms were currently part of a co-op, but the co-op's future was uncertain because only a handful of the farms were doing well. In another community the co-op had gone under because of competition from a co-op in a larger town.

One difficulty was that some of the co-ops had expanded too rapidly and spread their activities too thin or over too wide an area. Another difficulty for some co-ops was being purchased by large for-profit corporations or being unable to compete with those companies. Apart from those difficulties, co-ops generally remained active and were among the communities' most important local entities.

Organizations that historically attracted merchants and professional people living in small towns were less common among the farmers we interviewed. Hardly any mentioned belonging to Rotary International, Kiwanis, Lions, or the local Chamber of Commerce. There were a few exceptions. For example, several towns of moderate size had specialized subcommittees of the Chamber of Commerce, including subcommittees dealing with agriculture or with agriculture-related issues such as wind energy or water policy. Other towns had heritage associations, booster clubs, economic development associations, and agritourism committees that included townspeople and farmers.

Formal organizations fulfilled several important functions that could not have been accomplished as easily by kin networks or informal relationships among friends and neighbors. These organizations conducted business and thus contributed to the economic well-being of the community. Co-op boards made decisions about investments and rebates, for example, and growers' associations promoted marketing.

The organizations also provided important geographic linkages. Most of the neighborly relationships farmers described were among their immediate neighbors who farmed within a few miles and did business at the same grain elevator or gas station. The formal organizations were more often countywide, bringing farmers together from several local communities, and representing more general interests.

Some of the formal organizations farmers mentioned played a direct role in the civic betterment of their communities. It was not just that farmers schmoozed at board meetings or cultivated personal ties with other farmers. The organizations conducted business on behalf of the community, figuring out how the community could keep from losing its school in one case and bringing in a farm implement dealer in another. Some of the farmers' co-ops sponsored little league teams and school activities. Some of them donated funds to maintain the local park.

The implicit function that formal organizations played was to *limit* the extent to which farmers had to interact with one another. The difficulty with neighbors, one farmer observed, is that they hang around and get in the way. People in cities and suburbs say the same thing. It may be better not to know your neighbors if you want to be free to come and go as you please.

"It used to be everybody knew the neighbors," a farmer in his seventies recalled. "We had card parties and barn dances and picnics." But now, he said, "It seems like our life has been planned for us." He had formal organizations in mind. Schools, churches, and farm organizations. He wasn't sure why, but he felt less close to his neighbors than he used to.

The formal organizations farmers said they belonged to seldom met more than once a month. Some met only once a quarter or twice a year. There may have been a meal or time to chat before and after the meeting, but usually the meeting had a program and was devoted to conducting business. A quiet person was not expected to say much. Besides that, the organizations that included elected officials usually had term limits. Unlike having to deal with neighbors for years on end, it was possible to get in and get out.

The nonfarm organizations in which farmers participated were usually churches and schools (churches are discussed in the next chapter). The schools were located in town. They had long since replaced the one- and two-room country schools that had served the communities in earlier days.

"The main socializing that we do would be at school functions," a woman whose children were in high school noted. Her husband said the farmwork seldom gives him much time for socializing. But they agreed that during halftime at football games or after a school meeting of some kind, "the guys get together and start talking farming."

A woman who farmed with her husband in the cotton belt made a similar observation. She said there was little interaction among the farmers in her community who held adjacent tracts of land. But they did get together because of church and school activities. "We all send our children to private white schools," she said, "so we have those social gatherings."

Another implicit role that formal organizations played was to channel competition. This was a rather novel idea, but some of the farmers we spoke with saw it as a helpful contribution. Even without the increasing competition they sensed in their communities, they figured farmers were always competing with one another in small ways. It might only be trying to plant rows in the straightest line or getting one's corn planted first. Or bragging at the co-op about the wheat harvest. "The yield contest," one farmer called it.

Competition of these kinds could be all in fun. But it could be all too serious when it related so closely to farmers' self-esteem. Organizations provided ways to compete that were truly just for fun. Although they might be taken seriously, community norms set them apart from real life. It was possible to joke about who won this year's barbeque contest or who won the pancake-flipping race. The county fair included shuffleboard competition as well as pie-baking contests. The co-op might sponsor a charity raffle. The 4-H club might compete with itself to clean up the park better than last year.

For all these reasons formal organizations were present and important in all the communities we studied. Farmers were seldom active enough to say that the organizations took much of their time. But that was the point. It was possible during slow times of the year to be involved and then to be uninvolved when things got busy. This was apparently one of the reasons that farmers were not very involved in town-oriented organizations. They ran on a different clock.

SHARING INFORMATION

We asked our interviewees where they get information when they are making day-to-day decisions about farming. Traditionally, this was one of the important reasons for neighborly conversations. Neighbors shared their impressions of whether the season would be dry or wet. Discussions of when to plant or how to save on fuel took place over coffee on rainy mornings.

Farmers acknowledged that they listen for information from neighbors they happen to see at the co-op. Some new piece of equipment comes along, and they talk to a neighbor who has one to see if he likes it.

"A lot of it is in the evenings," a young cotton-belt farmer remarked. He talks to a neighbor in his seventies at the end of the day when they are out checking their irrigation lines. "I'm new to agriculture. I don't have all the answers," he explained. "So I spend a lot of time asking the older farmers questions."

"You're out talking with somebody," a farmer in a small wheat-growing community said. "And they'll say so-and-so did this and such-and-such did that. You always want to know what people are doing. You want information. You want to learn what's happening." He pauses for a moment and adds, "I don't know if that's good or bad, but that's the way it is."

One reason it was hard to know whether information was good or bad was that farmers were reluctant to acknowledge their failures. If things went well, they said so. If things went badly, they remained quiet. That did not stop gossip from spreading. It simply meant that the stories were incomplete. As one community leader noted, "You get all these rumors going around because the person who could give you the full-fledged answer isn't talking."

The information shared in these ways was often enough wrong or insufficient that agricultural policy experts sought to intervene in local networks. County extension agents hosted informational meetings or visited individually with farmers. Experiment stations set up test plots and circulated information about the results.[11]

With information increasingly available through the Internet, the possibilities for farmers to circumvent informal networks among neighbors have of course grown. Sharing over coffee on rainy mornings may be less important for this reason. Especially when specialized scientific and technological innovations are involved, the relevant sources of information are likely to necessitate different kinds of social networks.[12]

One of the women we spoke with said it was still common among the truck farmers in her community to learn about new seed varieties or chemicals from neighboring farmers. The reason was that the seed or chemical company would ask one of the farmers to host an open house. Then a representative from the company would come and everyone would sit around and drink coffee and eat donuts while they discussed the new product.

But that pattern was rare among the farmers we interviewed. The more typical practice was to seek information individually from specialists, or to do so online or at community-wide meetings, rather than to rely as much on neighbors. At least informal conversations with neighbors depended heavily on trust and were subject to being discounted on grounds that gossip was often inaccurate.

For example, Mr. and Mrs. Bower get their information mostly from the agricultural university and its county extension service in their state. He says a couple of his neighbors are pretty knowledgeable about agronomy, so he sometimes talks with them about seeds and fertilizer. But as a rule he doesn't. "I don't get that much information from neighbors," he says, "and what information I get, I tend to discount because if you ask someone what the best way to farm is, they respond by saying, 'What I'm doing is the perfect way.'" He wants "objective information" instead of neighborly advice.

The Rayburns feel the same way. Although Mrs. Rayburn's job in town gives her opportunities to talk with farmers, those conversations are professional and she does not share the information when she talks socially with friends. Mr. Rayburn prefers information from the county extension agent or from the company he uses for seed and fertilizer. He has some research plots to test different seed varieties. That gives him firsthand information, rather than relying on talking with neighbors.

His conversations with neighbors have to be taken with a grain of salt. "I can get an inch of rain," Mr. Rayburn says, "and if I tell that at the coffee shop, the next person will say they got an inch and a half." By the end of the conversation nobody knows for sure how much rain anyone got.

Farmers' preferences for specialized information have risen significantly, they say, because of technologically advanced equipment and commodities. They knew there was a lot more to understand about hybrid seed varieties, new chemicals, soil science, and animal breeding than could be communicated along the grapevine.

The woman whose fellow truck farmers learned about new products over coffee and donuts provided an interesting illustration of the changes. She lives an hour's drive from a research lab run by the agricultural college in her state. When a question about a crop disease pops up, she drives to the lab to discuss the problem in person. She also hires an independent lab to conduct routine soil tests.

A farmer in his late forties who operates a 2,500-acre corn and soybean farm provided another illustration of how technological innovation affects information sharing. He never checks with neighbors for technical advice. "We really do care about the people around us," he says, adding that he wants to maintain good relations with them. But that is difficult because he raises pigs and the neighbors complain about the smell. He hires an environmental specialist to make sure he is not violating regulations.

Farming on that scale makes it possible to consult additional sources of special expertise. He has a tax accountant to do his taxes and help with estate planning. He consults his financial advisor daily about commodity prices. He has a futures app on his smartphone and a satellite dish atop his farm office. He checks the weather app when he is spraying because the spray is good only for an hour and is wasted if it rains too soon. He has an independent crop consultant. The consultant comes out weekly, drives an ATV around the fields, and reports on weeds and insects needing to be controlled.

Information sharing was changing for other reasons as well. One of the farmers we spoke with had met his wife on the Internet. He was the only one he knew who had. But it was the kind of example that illustrated how rural friendships and even marriage markets might be changing.

As they looked to the future, farmers could imagine trends and countertrends simultaneously. As farming communities became smaller, a kind of social implosion could be imagined in which the remaining farmers interacted more closely with one another than ever before. That might especially be true in communities where the remaining farmers were interrelated. The countertrend implied that farmers would look beyond their immediate communities for social relationships relevant to farming. The Internet, travel, and connections with universities and business contacts in cities illustrated these wider networks.

FARM TOWNS

Besides the fact that farmers were busy farming more land and using new technologies that required more specialized information, the biggest change in farming communities was that many of the local market towns were struggling. True, there were exceptions. Several of the farming communities we studied were within easy commuting distance of medium-sized cities. But most of the towns in the areas we studied were small, and many of these towns were losing population.[13]

This was not entirely a new phenomenon. The population of small towns in rural communities peaked in many areas before or just after World War II and has been declining ever since. That decline continued during the farm crisis of the 1980s and was still occurring during the first years of the twenty-first century. It was especially acute in areas where the farm population was also declining, in towns that were already small, and in municipalities that did not benefit from being a county seat, hosting a college or junior college, or having an interstate highway nearby or a city within convenient commuting distance.[14]

Some of the larger towns were faring better than might otherwise have been expected because retired farmers were moving to town or because active farmers with large scattered acreage were living in town. But those demographic shifts seldom made up for the fact that fewer people were farming more land.

Towns composed of retirees and counties with aging farmers necessarily had smaller proportions of children. That affected the community's social life. It was harder to keep the school open than the nursing home. A farmer we spoke with in a county of thirty-five hundred provided an interesting example. Seventy percent of the county's population was retired or semiretired. There used to be fifty students in a class. Now there were ten. The town was in danger of losing its school. If the school closed, the man said, it would be "like losing the last fifty years of your little town."

The demographic shifts were only part of the story, though. Small towns still had the occasional grocery store and filling station, a school, and perhaps a grain elevator or farm machinery and repair service. But more of those business activities were being conducted out of town in large places by chain stores such as Walmart and Home Depot.[15]

Like town residents, farmers marked their towns' trajectory less by the gradual decline in population than by the loss of a school or an essential business. They could do without a clothing store, but needed a hardware store, a repair shop, and an implement dealer. Many of the farmers we spoke with were angry with implement dealers such as John Deere and International Harvester. They said the companies were so intent on making money that they cut off the dealers in smaller towns and forced farmers to drive twenty or thirty miles for parts and repairs.

A corn-belt farmer offered some numbers he had heard recently to illustrate the change in equipment dealers. In the late 1970s more than 700 dealers belonged to the farm equipment association in his state. Since that time, the association had merged with one in a neighboring state. Now there were fewer than 300 dealers in the two states combined. On average the commute from farm to nearest dealer in the 1970s was nine miles. Now it was twenty-one.

Farming towns were also declining, farmers said, because a smaller share of farm income was staying in the local community. It used to be, they said, that a dollar from local agriculture circulated in the local community seven or eight times. It went to the local co-op to pay for seed and fuel, to the county treasurer to pay for teachers' salaries, and to the local bank, grocery store, doctor, and barber shop. The money no longer circulated locally if the school no longer existed and the doctor was in a larger town thirty miles away.

They also thought more of the money was leaving the community because of large businesses playing a more important role in agriculture. Seed was no longer produced by local farmers but by a large chemical company that invested its profits at best in cities or at worst in offshore tax havens. That put local businesses at a disadvantage.

To protect local businesses farmers and town residents said they tried as often as possible to shop locally. In one of the communities we visited several dozen farm families had each contributed several thousand dollars to keep the local grocery store from closing. It was like buying shares at the farmers' co-op, except they did not expect to receive dividends in return.

It was more evident, though, that farmers were increasingly making major purchases outside the local community. Seed was cheaper purchased in bulk and trucked in from another county or state. Machinery might be purchased online and brought in from a distance as well. Insurance and brokerage services were located in larger towns. When practicality conflicted with sentiment, practicality won. "We lose local businesses all the time," one farmer acknowledged. He added, "But hey, that's life."

WORK AND SOCIAL LIFE

Although the demographic changes were having a significant impact on community life, the farmers we spoke with were more concerned that the neighborly relationships they valued might be eroding for a different reason. The pace of life was simply faster. Despite laborsaving equipment, the work was no less demanding. In a dog-eat-dog world, everyone was working harder and longer to keep up. Or at least it seemed that way. The actual number of hours working per year might not have changed. The physical labor was probably less difficult. But the temptation was there to focus all one's attention on work.

Farmers who grew up in the 1950s and 1960s associated the recent demands with what they perceived as more relaxed times in those earlier decades. They recalled families getting together on Saturday evenings to play cards or eat dinner together. Men went to the Legion hall or the local pub and drank beer. Cousins got together for birthday parties. Those were happier times, slower times, at least in memory. Time for socializing now seemed rare. It was not that people cared less about their neighbors. But it was harder to make time for them. Evenings that would have been spent with friends and neighbors were taken with filing farm reports and keeping tabs on grain prices.

The farmwomen we spoke with felt this tension more acutely than many of the men. They said the men were off somewhere harvesting or planting or tilling the fields instead of being available to attend a ball game or a school play or have dinner with friends. The men agreed that it was often difficult to be available or to schedule any time to socialize with the neighbors.

"Don't let your work run your whole life," one woman counseled. "Take time to enjoy other things than your work." She said the tendency among farmers is always to say, "Oh, we can't do this or that because we've got to get this done." She warned, "You can end up, if you aren't careful, not having any kind of a social life other than farming." She knew that from personal experience. "That's what I'm saying because that's been our life."

Her warning was one that a farmer in his forties who was farming twice as many acres as his father agreed was worth taking to heart. Although there were never enough hours in the day, he argued that one of the real pleasures of living in a farming area involved taking time to eat breakfast with friends in the community. He thought laborsaving equipment and fewer farmers tending livestock should, if anything, increase the opportunities to do this.

This farmer considered it especially valuable to interact with older farmers. Most of the active farmers in his community were in their late fifties or sixties. Many of the others were semiretired or retired. The interaction between generations is priceless, he said. His whole adult life he has eaten breakfast with seventy-year-olds at least once or twice a week. When he was a young farmer, he mused, nobody told him that he would meet "twenty of the most wonderful people in the world" through farming—or that he would go to all their funerals.

On the whole it was clear that farmers still thought that living among friendly neighbors was one of the best things about farming—and that they were worried about the declining number of neighbors, the competitiveness, and the busy schedules that prevented them from enjoying their neighbors as much as they would have liked. But those concerns did not imply that farming communities were collapsing. Nobody we spoke with thought the ways in which farming is currently done are so fundamentally wrong that something vital needed to change to restore the vitality to their communities. They were more realistic than that. They understood that neighborly relations can be maintained without spending huge amounts of time socializing. They also knew that social relations are complex, diverse, and increasingly dispersed.

71

FAITH

3

My early mornings when I'm sitting there with no noises around, well, that's kind of my time and God's time together.

—Cotton-belt farmer, male, age 34

Faith is important. I'm not part of an organized religion. But I can't say I don't think about God every day.

—Wheat-belt farmer, female, age 50

Do people whose lives are so deeply influenced by the weather and by other forces beyond their control find solace in faith? Scholars have long considered it likely that people whose livelihoods were so powerfully shaped by the elements would imagine that those influences were somehow divine. Why else would peasants in medieval Europe have held special religious rituals at the moment when spring planting was being done or when farm animals were giving birth?

The history of religious practices in the United States, though, suggests a more complicated pattern. To be sure, there were churches in abundance in rural communities. Farmers flocked to revival meetings, where they promised to repent of their sins in hope of receiving God's blessings. They prayed for rain and sang hymns about bringing in the sheaves. At the same time closer inspection suggested that farmers were often too busy to attend religious services faithfully and that, if anything, church participation increased as less of the population lived on farms and more of it was located in towns and cities.[1]

In designing the Princeton study, I wondered which of these traditional views of religion in farming communities would be more prevalent today. On the one hand, farming is still a business shaped by the weather. An entire year's work can come to naught if the rain is too little or too late. Farming is a dangerous occupation that involves risks from accidents as well. And as an aging segment of the population, farmers are subject to higher than average rates of illness, bereavement, and death.

Those factors would all suggest that religious faith is vitally important in farmers' lives. Were those not reason enough, religious involvement would also seem likely because of the intergenerational ties among farm families and perhaps even because of being located in small out-of-the-way places with fewer alternatives for leisure and entertainment.

On the other hand, the demographic and technological changes taking place in farm communities seem likely to be affecting religious habits as well. It would seem to matter that farm communities are losing population and that technology is shaping how farmers spend their days.

Ultimately the question of how faith is faring in farm communities is not an either-or proposition. We spoke with farmers who said their religious faith was central to their lives. We also spoke with farmers who had little use for religious faith. The important aspect of these conversations was *how* religious faith was meaningful and *why* it was or was not meaningful. The language farmers use in talking about religious faith is a window into their lives. It sheds light on the risks they experience from day to day and on the constraints that are part of their daily activities.[2]

Nearly all of the farmers we spoke with had experienced setbacks in their work and tragic times in their lives. Faith in God mattered in these times. It usually mattered because farmers had a kind of nominal, ongoing sense that God existed, rather than from having spent a great deal of time cultivating that relationship. Farmers' piety in this respect appears less as a matter of deeply detailed creedal convictions and more as an aspect of the taken-for-granted realities of life.

Church going for the farmers we spoke with was rather different from their implicit understandings of God. Although they listened to sermons at church and sometimes participated in prayer meetings and Bible study groups, these activities were built into the normal round of social activities that farmers were expected to support in their communities. Church participation was in this respect similar to attending meetings of the farmers' co-op or being present at the high school's athletic events. As was the case with informal relationships among neighbors, the ill effects of conflict influenced church involvement. When conflicts occurred it was difficult for farmers who expected to live all the remaining years of their lives in the same community to know how to respond.

The particular wrinkles associated with faith and religious participation in farming communities were evident to the clergy we spoke with as well. Many of the clergy were themselves from farm families and were perceptive observers of life in rural communities. At the same time clergy interacted most often with the few people in their congregations who happened to hold leadership positions or were active in small fellowship groups. They knew instinctively that those people provided a skewed view of the larger community. They probably would have been surprised, though, to know the extent to which some farmers' held negative views of religion.

FAITH AS DIVINE SUPPORT

The farmers we spoke with rarely were the kinds of people who wore their religion on their sleeves. They talked more openly and at greater length about other topics and seldom made reference to God or faith in discussing those topics. But when asked directly about religion, they generally had comments and experiences to offer. Like other Americans, they mostly expressed belief that God exists and refrained from delving very deeply into theological nuances.

The older Jorgensens were typical in this regard. Clay said his wife could probably talk about it better than he could, but in his opinion the best way to live was to "just let the good Lord handle it" and then a person could relax and figure things would turn out okay. Mary said the good Lord had seen them through some pretty difficult times. "You know, as far as farming and God are concerned, I always say that God is our boss. We have to work with Him every day."

Neil Jorgensen saw things a little differently but agreed that religious faith was the key to dealing with life's "ups and downs." He thought that praying and going to church and having faith in God were pretty firmly rooted among the farm families in his community. It certainly was the case on his mother's side of the family, he said.

Other farmers agreed that farming might be the kind of occupation that encourages people to believe in God. "Farmers have an awareness of their lack of control" was how one farmer put it. "You've got to trust in God," another farmer noted in discussing what he learned growing up on a farm. "You do the best you can," he recalled his mother telling him, "and there's no need to worry about things you can't control."

A farmer who attended a Methodist church in the cotton belt expressed the point more graphically. "God's got a way of knocking us back on our butt," he said. "When you get to thinking you're running the show, that's when you've got a problem. He's got a way of saying, hey, I'll show you who's running the show."

They thought, too, that maybe the Bible was especially meaningful because so many of the stories were about farming. Farmers "understand all those parables in the Bible," one woman noted, referring to the fact that some of the parables described people working in the fields, gleaning wheat, or tending sheep. A man who had been farming for four decades explained that religion "plays a key part" in farmers' lives "because we're confronted with creation every day—you know, you see the miracle in the seed every season. You just can't hide from a big picture when you're a farmer."

Many of the farmers we spoke with said their faith in God sustained them during especially difficult times. Drought, hailstorms, crop failures, livestock dying, injuries in farm accidents—all these were such times. It was not surprising that people of faith turned to God during these times. *How* they felt God responded was the interesting part.

Ruth and Larry Pfluhoft are third-generation wheat farmers. They are in their mid-fifties and are now farming about fifteen hundred acres of land. Half of it is rented, and they are making payments on the part they own. Mr. Pfluhoft supplements the farm income doing construction work during the winter months. They are people of faith who attend worship services at least once a week. Sometimes more.

Nearly everyone in their community goes to church. About half are Catholics. The others are conservative Lutherans, Methodists, and Mennonites. A few commute some distance to a fundamentalist church that has been growing lately. Religion is such an important part of their community that nobody does fieldwork on Sundays. The Pfluhofts are Catholics. He serves on the parish council. She sings in the choir and teaches one of the children's Sunday school classes.

When the Pfluhofts got married, their dream was to operate a dairy farm. Mr. Pfluhoft's parents and grandparents were dairy farmers, and he wanted to follow in their footsteps. Mrs. Pfluhoft loved working with animals too. Her parents and grandparents raised wheat and had chickens and sheep. But the dairy business did not work out. Mrs. Pfluhoft recalls the difficulties that proved the idea was just not feasible. Mr. Pfluhoft chimes in. "I got some good guidance from the Lord," he says.

Later in the conversation, the topic turns to a recent season too dry to even plant the wheat. Eventually it rained, but by then it was too late. When the Pfluhofts have family rosary, they often say "in thanksgiving for rain." It reminds them that God is in charge.

"There are times we think, 'Lord, you made a mistake,'" Mr. Pfluhoft says. Mrs. Pfluhoft adds, "We feel we should be in control and God has a way of saying, 'Ha-ha. Fooled you on that one.'" They think having faith means believing that God is ultimately in control. "We thrive on being independent," she says, "but then God's in charge. Our faith tells us that we've got to trust in God. It always works out."

The Pfluhofts' understanding of how faith works was shared by many of the farmers we spoke with, Protestants and Catholics alike. It was reassuring to believe in God when crops failed and cattle died. The idea was not that God would prevent these things from happening. "It'll always rain in time for God's crops," was how one farmer put it. Rain did not come because people prayed for it. It was rather that God was a kind of constant in life, a background source of stability. A person of faith gave thanks for rain whether it rained or not. Difficulties were survivable because God was ultimately in charge and would work things out.

"There's always a purpose for whatever happens" was how a cotton-belt farmer in his fifties put it. He and his wife went to the Baptist church. Neither had been to college. They spoke with a strong Southern accent. "Why these droughts come, I have no idea," he said. "But I know who holds the reins. I know he don't make mistakes. Maybe it's to get our reaction on how we're going to treat our fellow man when these things come along."

The logic evident in these and similar remarks is comparable to the view of God that has been identified in other studies of contemporary religion. Even though a person prays to God or works hard in hopes of pleasing God, the idea is not that what a person does actually causes God to respond in a certain way. A farmer may hope that prayer will bring rain, but the exact nature of that hope has less to do with rain than with being aware of God's existence and thankful for God's presence, no matter what.

"We sort of laugh at each other in the spring when we have to start watering," a farmer who described himself as a born-again Christian noted. "Well, don't you have faith it's going to rain," someone will say. "Yeah, I do have faith it's going to rain," he replies. "But the good Lord gave me the ability to know when to irrigate when I need to irrigate. I don't sit on my can and wait for him to do it for me."

His view was a bit like the popular notion that God helps those who help themselves. That was not quite what he was saying, though. It was rather that faith in God needed to involve neither superstition nor a belief in magic to be real. There was a message in his willingness to joke about God making it rain. For him God was there whether it rained or not.

That was how a truck farmer in his forties who attends the large Catholic church in his community saw his relationship to God as well. He said there were some especially stressful times of year, such as picking season, when he has to deal with more employees than usual and when safety is more important than ever. He says he makes it a special point to attend mass and pray when he knows those times are coming. It isn't so that God will intervene, though. Instead, his faith helps him "decide what's the best way to handle it."

Some of the most difficult times farmers described involved illnesses and the death of loved ones. Scripture, prayer, words of comfort and support from clergy and friends provided sustenance. The words conveyed added meaning because they intertwined with object lessons gained from farming. Death and renewal were part of the cycle of life, evident in the seasons and the crops.

An older farmer explained how his faith had sustained him a few years earlier when his wife died. "Things will never be the same," people said, and he knew that was true. He also came to understand from reading scripture and from thinking about what he knew from farming "that out of chaos and out of very difficult times can come very wonderful things, in fact, beautiful things." He felt his life was proof that miracles can happen. He meant the miracle of understanding the beauty and wonder of life even though his wife had died.

But the farmers we interviewed were reluctant to say that farming and religious faith were related in some unique or distinctly powerful way. Faith may have been especially meaningful to them when a family member died or a crop failed, but they noted that everyone—farmers and nonfarmers alike—experience difficult times and may be assisted by faith on those occasions. They may have prayed for rain or given thanks for a bountiful harvest, but they did not want to appear superstitious. The rain came or did not come whether they prayed or not.

Had they linked concrete results more closely with prayer, they likely would have been disappointed, perhaps to the point of questioning their faith. One of the older couples we interviewed gave a poignant example. They said it probably was more common for people of their generation to pray for rain and health than it was among younger people. That was the context in which they and their friends had prayed fervently for the doctors to cure their son's brain tumor. Eleven months later, they buried their son. "I'm just at a bitter time now," the man said. He felt that way because he had fully expected God to heal the tumor, not simply to be there to strengthen the family when their son died.

Although it was more salient in times of crisis, faith for most of the people we spoke with was a kind of constant, a feature of life, a part of one's worldview that depended less on heavy-duty theological interpretation and more on simple statements of belief. "Faith is more 'love God, love neighbor,'" one farmer explained, than anything else. "The transitory things such as whether it is going to rain this year or not or am I going to be earning my living from the land ten years from now are secondary. Faith transcends the crisis of the moment."

A wheat-belt farmer in his fifties who regularly attended church with his wife said his faith was important because it was "always going to be there" and was "something I can always count on." He meant that God was always there, unlike his rented farm ground, which he knew might not be. It was his faith, though, that was a constant in his life, as much as it was God. That was literally true. He grew up in the same Baptist church he was still attending.

Mr. Rayburn took a somewhat different approach in explaining how faith was a constant in his life. Living as he did in the cotton belt, he thought religion was definitely important to most of the farmers he knew. They were mostly Baptists or Methodists, and the recent immigrants in the area were

Hispanics who were either Catholic or Pentecostal. He and Mrs. Rayburn went to church just about every week, and they served on church committees. He was reluctant, though, to think that God was somehow willing to intercede actively in daily affairs. Reluctant, and yet not entirely willing to rule it out, either.

"Why don't you pray about it?" Mr. Rayburn says this is what his wife tells him, almost as if she is nagging him to believe more in God than he does. "I pray every day," he tells her. "I don't see you," she replies. "I do it my own way," he explains. "She might be more outspoken about it, but I have good contact with the Lord. I just keep it private to myself."

"I'm not a holy roller," he adds. He prays, but prefers to keep it to himself. Just the other evening he was at his daughter's softball game. "I said a little prayer to myself. I said, 'Lord, let her hit the ball.'" Sometimes she did and sometimes she didn't. It didn't matter. The point was not that God would intervene. It was rather that prayer was an ordinary part of life. Praying was a way of staying in contact with the Lord.

FARMING GOD'S CREATION

The idea that God is always present, at least in the background, was reinforced among the farmers we spoke with by their sense of being in direct daily contact with God's creation. They experienced the rhythm of the seasons. They saw plants and animals come to life, grow, mature, and die. Doing fieldwork positioned them to witness the changing cloud formations. They saw the sun rise and saw it set, felt the wind, and knew the smell and composition of the soil. Larger and more powerful equipment notwithstanding, they felt a connection with the created order.

It would have shocked the farmers we spoke with to suggest that they were part of a long American tradition of naturalistic spirituality. They were not the latest incarnation of the Transcendentalists, of Ralph Waldo Emerson or John Muir. They were more a part of mainstream Christianity than that. And their relationship to the land was less passive and more instrumental. But they did understand the close relationship of God and nature.[3]

They were more aware of this connection when things went wrong than at other times. Nothing demonstrated that they were not in charge quite as clearly as an untimely hailstorm. An event of that kind conjured up images of a wrathful God. Or at least an unpredictable God.

Studies of other sectors of the American public, though, suggest that wrathful, unpredictable deities may be losing popularity. That kind of God appears to be less believable than one who behaves with greater consideration. Americans prefer a temperate, friendly God.

Farmers are no exception. The benevolent God farmers believe in is a creator deity who makes things grow. Working the soil, they are reminded of the creator's presence. Being outside, close to nature, they feel a special connection to God's creation. It is an enduring connection they know is there whether they pause to think about it much or not.

"You plant the crop, and if He doesn't help you," a wheat farmer in his sixties says, "you don't get anything. He's got to stay with you through that whole crop to make it work."

The farmer pauses to reflect on what he has just said. He figures it is important to pray and ask God for guidance. A farmer should do that, but so should someone working for a paycheck in town. So what is the difference, if any?

"Farming is 100 percent," he decides. The crops, the weather, a hailstorm, the rain, the livestock, a snowstorm that kills the calves—it's all related. "Getting your paycheck is through the Lord," he says. "I don't know how to explain it, but He needs to be with you every day to make it work."

That feeling of dependence, notably for some of the farmers we spoke with, was also an occasion for something bordering on celebration. They knew they were not in charge. They understood that they were vulnerable. But it was reassuring to know that God had things under control. More than that, it was a privilege to work where God's involvement was so fully evident.

"The part of farming that is especially nice," another wheat-belt farmer explained, "is that you get to see how Creator God has made our ecosystem and how all that works together." She said one of the most enjoyable aspects of farming is "the opportunity to see what God has made and how these things all relate."

A cotton-belt farmer spoke in similar terms. "You cannot be a farmer and not believe in God," he asserted. "I just see his hand in the farming community every day in the way the crops grow. There's no getting around it. When you go to the field in the morning, you see the sun come up, you see the sun go down in the afternoon. You see the hand of Providence in everything."

A related view is that farming is a particularly worthwhile endeavor because it fulfills a divine mandate. The mandate is not one of subduing the earth. It is rather to produce food. Helping feed and clothe the world is an ethical responsibility grounded in biblical faith.

The land in this view is almost sacred. Almost but not quite. There is a special obligation to take care of the land, another farmer observes. Being a good steward of the land means not only making it profitable but also respecting its relationship to God. Stewardship is ultimately toward God. The land is God's creation and thus special, but not to be confused with the creator.

Distinguishing the creation from the creator is more a practical matter than an abstract theological tenet. It permits the land to be treated in practical terms rather than worshipped. The land is constant and enduring and yet variable in what it yields. God is more constant and enduring and invariable than the land. This is the "bigger picture," farmers say, that undergirds their faith and reduces their anxiety.

"We really truly believe that this land belongs at the end of the day to a higher power than we are, to God," a woman who attends a rural interdenominational church near the six thousand acres she and her husband farm in the corn belt says. Knowing that reminds her to be grateful rather than taking too much of the credit when things are going well.

The distinction between creation and creator puts the land in perspective in relation to other values as well. A farmer who attended an independent evangelical congregation composed mostly of farmers captured the idea clearly in explaining that the land was only a part of God's creation. "Your day-to-day contact with people, your relationships, that type of thing is as much or more important than the land itself."

PARTICIPATION IN CONGREGATIONS

In the farming communities we visited we heard different perceptions of how the churches were faring. The differences partly reflected variations in population trends. In areas where farms were larger and fewer, the church-going public was generally declining as a result. In other areas where the population was stable, church participation was doing better.[4]

The explanations people give for social trends are often as interesting as the trends themselves. In the case of church going, an argument in the popular press and in many religious circles is that declining participation is an indication that something terribly wrong is happening. America is turning away from God, becoming secular, selfish, and materialistic. Or if participation is increasing, that is sure evidence of good preaching or people turning to God.

We found farmers voicing rather mixed views about the meaning of religious decline. On the one hand, it was evident enough that churches in their area were declining because the farm population was declining. On the other hand, they were still concerned that churches were faring badly because of something more worrisome. Even if they personally were not as active as they used to be, they linked declining church participation with trends they considered disturbing.

"Churches are closing up," one woman observed. "That's what [is happening in] these small towns where the larger farmers are coming in." "It isn't good," her husband echoed. They did not farm a large amount of land.

Throughout the interview they expressed concerns about land in the area being taken over by large farmers. Church closings were further evidence that this was a problem.

A different concern was that even the people who still lived in the community were no longer supporting the churches. "Young people are falling away from a faith in God" was how one of our interviewees put it. Another farm parent worried that "young people are starting a lot more to walk away from church." He thought that was not good. At the same time he felt that sports and school activities were taking up so much time that young people were busy doing other things.

Besides having other things to do, younger people in farming communities were increasingly experiencing the fact that fewer of the people at their churches were their own age. That was especially true among teenagers. It was also true of younger farmers. It was hard to stay involved in a church that had no activities for people their age. Or for families with children.

"It's kind of surprising to see a young person in church," a corn-belt farmer in his late seventies remarked. He and his wife are regulars at the Lutheran Church close to their farm. "You just don't find them in church anymore," he said. "The old duffers are there." "Fifth row from the front," his wife chuckled. She and her husband always sat there. That was their spot.

"I go to a men's prayer breakfast on Thursdays," a wheat-belt farmer in his forties who attends a Methodist church related. "I look around the room and there are eight of us there. In ten years there will only be two probably. Only two of us are under the age of sixty. In ten years it could just be the preacher and me. The rest of them will be gone." He had no doubts that the whole congregation would be smaller in ten years.

One of his neighbors had recently turned sixty. His two daughters and their husbands farmed just up the road. They rarely went to church. He blamed himself. "My girls don't go as often as I wish they would," he reflected. "When they were young, I didn't make it important enough." As a young farmer, he felt he had neither the time nor the energy. "We had to make a living," he said. "We were trying to chase the bankers away."

Not everyone who felt the churches were declining decried the decline. On the whole the farmers we spoke with were realists about the amount of commitment the churches deserved. They were quick to assert that they were not fanatics. They were Methodists, Lutherans, Catholics, or Baptists. They had been all their lives. There was a time when everyone was expected to hitch up the horse and buggy and get to church even when it was cold and snowy. They were mostly glad that those days were over. It made sense, some said, to call off church from time to time.

"I see a lot of hypocrites in church," a farmer who felt it was good to call off church when the weather was bad explained. He is all for people having a good moral upbringing and believing that there is a Supreme Being. He goes to the Lutheran church in his town, but worries that a lot of people go for fear of what people will say if they don't.

Perceptions also varied depending on farmers' own experience with local churches. Some of the farmers we spoke with felt that farmers still had an ethic of joining and helping that contributed vitality to local churches. As one farmer put it, farmers in his community were "the active people in the church putting on the church suppers, working in the fire station, and running the ambulance." He thought it was "their nature to just band together and try to get the job done." Others' experience left them doubtful.

The problems that led people to doubt the vitality of their own congregations included fewer members, an aging clientele, and difficulties in attracting and supporting a pastor. Just as among residents of small towns, farm families complained that their church had actually closed or was having services less often or did not have a pastor. Those changes made it more difficult for them to stay interested and involved. They used to go to St. John's where they knew everybody, but now they had to drive twelve miles to St. Patrick's. It took more energy to go and when they got there it did not feel like home.

These difficulties notwithstanding, many of the farmers we interviewed saw their church as the center of their family's social life. They attended as regularly as they could except during harvest or when planting had to be done or when they were away on vacation. They served on the vestry or deacon board. They helped keep the building in good repair and brought casseroles to church suppers.

The principal theme in these activities was that they were *practical*. Few of them had anything to do with doctrine, biblical teaching, or worship. They were instead routine activities concerned with maintaining the building, keeping the grounds clean, serving on committees, and organizing potluck dinners. "Everything from mashing potatoes to installing new insulation to chasing the bats in the belfry," one woman explained.

At this woman's church one of the more interesting routine activities was killing the pigeons that insisted on making the church their home. It was a grisly business, whacking them with shovels, but it had to be done. Golden pigeon awards were given to the members willing to take on the task.

A truck farming couple offered a more palatable example. They said the church was their "main social activity." They were in their sixties and liked to get away on weekends to attend their grandsons' wrestling meets and football games, but they usually got back for Sunday services and were available to do church work during the week. Once a month she organizes a dinner at the church. From time to time they do more. They recently redecorated the parsonage from stem to stern. Painting, staining the woodwork, putting in new carpet—they redid the whole inside of the house.

Other examples ranged from planting flowers around the church to helping with the nursery to videotaping the Sunday sermons and putting them
on the church website. The annual highlight for one farmer was inviting the
congregation to his farm for an old-style barbeque. Other activities included
helping elderly members with home maintenance and auto repairs as well as
visiting the sick and preparing meals.

The various churches that vied for members in sparsely populated farming communities differentiated themselves through these routine activities
as much as through different styles of worship. Bingo was best at the Catholic church. The Methodists put on the annual pancake feed. The Lutherans
hosted the yearly fish fry. The differences were sometimes the occasion for
lighthearted remarks. "We're Brethren," a woman explained, "just an off-
shoot of the Baptists." She laughed. "We don't use Tupperware for covered
dishes. Ours are glass."

The routine activities farmers described reflected the important *social*
role that congregations play in farming communities. People expected their
congregation to host a monthly potluck dinner and to have a barbeque each
summer. That was the norm. The building had to be maintained. A farmer
with a front loader could pitch in repairing the sidewalk. A farmer with
good business skills could help with the church budget.

It was not that preaching and teaching were unimportant. But in small
congregations, the clergy handled those. Lay members expected the clergy
to shoulder these tasks. The implicit bargain included an element of defer-
ence to the clergy's special expertise in those matters as well.

The few lay activities that dealt specifically with congregations' theological emphasis were teaching Sunday school and hosting Bible study groups.
Among the farmers we spoke with, women more often than men did these.
Several of the women also served on regional denominational boards and a
few had gone on short-term mission trips to other countries.

The churches also functioned as families' mainstay when illness, death,
and tragedy happened. The most familiar role was to pitch in whenever a fu-
neral occurred. Besides the obvious fact that rural funerals were usually held
in church buildings and burials occurred in church-associated cemeteries,
church members organized the social activities surrounding funerals. These
activities usually involved food. Bereaved families would find neighbors
at their door with casseroles and homemade pies. Church kitchens would
serve meals before and after funerals.

Such activities reflect the fact that farm families usually live in their communities over an extended period and know the local customs. This is more
important than actually being close friends with one another. If friendship
was the key, funerals and the food-sharing activities associated with them
would be much smaller. Instead, people attend funerals and bring food to
families they hardly know, except for the fact that they live in the area.

Beyond these symbolic expressions of caring church relationships are sometimes the sources of more significant kinds of assistance. In one community, for example, farmers were going under because of poor crops that were making it hard to make the payments on farm loans. A men's fellowship group at one of the local churches discussed the problem and worked out a behind-the-scenes arrangement to provide temporary assistance.

Prayer chains were a more common way of organizing assistance. "Whenever anybody in the community calls and says they need prayers," a woman in a cotton-belt area explained, "we get on the phone and call everyone and start a prayer chain." She did not say that the prayer chain resulted in divine miracles, although that may have been the case. Instead, she said it was "comforting" to know that "lots of people are praying for you." That knowledge mattered in itself. The prayer chain also formed the basis for providing assistance, such as driving an elderly person to the hospital.

Besides their role in helping individual families, rural churches were sometimes a pivotal organization in dealing with community-wide crises. One of the farming communities we studied had experienced severe flooding within the previous year. Some of the farmland closest to the river was under nine feet of water. Many of the stores and houses in town were damaged.

The churches organized a committee to enlist volunteers to do cleanup work. With fields too wet to do fieldwork, farmers helped repair damaged buildings in town. Their machinery came in handy for clearing debris. Their mechanical skills proved valuable to the physical labor involved. The churches also housed and fed some of the people whose homes were lost. The churches "kind of turned into the center of the recovery effort," one farmer recalled.

The main thing about rural churches is the organizational structure they provide in farming communities. Like farmers' co-ops, Masonic lodges, and other local organizations, churches give people who are otherwise too busy with their own activities a reason to come together. Believing that God cares about these activities is an important part of the equation. But so is the fact that the church roof periodically needs repair. Even the smallest churches have committees to maintain the building and committees to perform other activities, ranging from taking meals to shut-ins to organizing quarterly church dinners. The key is to structure these activities to take some of people's time without becoming a burden.

Like the local co-op, the Grange, the Home Demonstration Unit, and various other farm organizations, churches forged bonds among people who otherwise may seldom have interacted. The bonds were sometimes fragile, farmers told us, because farmers' social ties so often ran in family circles rather than extending more broadly. But churches were among the local organizations best positioned to bridge between farmers and townspeople.

Usually the churches were located in town rather than in the countryside and included retired farmers who lived in town as well as shopkeepers and farmers.

Like other local organizations, the churches functioned in ways that typically reflected and sometimes reinforced divisions within the community. Although there were times when different congregations joined forces, held joint worship services or holiday celebrations, and pitched in when community crises occurred, they functioned most of the time as separate organizations. Some of the time they represented not only different styles of worship and different doctrinal teachings but also different family histories and ethnic traditions.

A truck farming community provided an example of religious organizations dividing rather than uniting the local population. The three dominant religious traditions were Catholics, evangelical Protestants, and Mormons. Members of each had historically criticized the others on doctrinal grounds. In recent years the division between Catholics and evangelical Protestants had intensified as more of the former were Hispanic and more of the latter were ultraconservative Republicans. Non-Mormons admired the Mormons for taking care of their own, but resented Mormons' dominating the local school and hospital boards.

The factors that mitigated such divisions in most of the farming communities we visited included greater denominational diversity as well as intermarriage and social networks that crossed denominational lines. As one cotton-belt farmer explained, "I'm a Baptist but my best friend is Catholic. We have a good working relationship. I respect him. He respects me. There are no ill feelings because you're a Catholic and I'm a Baptist. We all get along."

The farmers we spoke with who said religious divisions were unimportant in their communities thought they once had been but were no longer because of newcomers and because the long-term families had learned to get along. They noted, too, that congregations were seldom composed predominantly of farmers. The more common pattern was for farmers to attend churches in town in which farmers were a small minority.

A kind of live-and-let-live attitude was evident as well. It did not imply an anything-goes approach to life. The view was rather that common moral values prevailed in farming communities—perhaps in nonfarming areas as well—that consisted of honesty, decency, and hard work and transcended religious traditions. It suggested that people could seek God via different traditions. *Which* tradition mattered less than the person's faith.

"There is only one God," a wheat-belt farmer observed. "I'm not going to condemn one religion or another. As long as you believe in God and try to live the way you think you should, you can look yourself in the mirror every morning."

Ironically, that view was shared by people who thought it *did* matter which church a person attended. They too believed what counted was a person's relationship with God. They just thought most of the people who attended churches other than their own did not have one.

"It's not just attending church," a couple who attended a theologically conservative nondenominational community church explained. "What's important is a personal relationship with the Lord," the man said. "We have to rely on the Lord to provide for us—and he has."

In their case theology mattered. But in a close farming community where everyone knew what everyone else was doing, it was also important for them to do things differently from their neighbors. "There are certain standards that the Lord requires of a Christian," the woman noted. "We have specific guidelines from the Bible," her husband added.

The most visible of those guidelines was refraining from doing farmwork on Sundays. "There are some in our neighborhood who work on Sunday," the woman said. "To us, that's wrong and that's something we don't do." "We see it as a day of rest," the man explained. The only thing they would do on Sunday is stop the dirt from blowing if it was blowing. Otherwise, they do not spray or harvest on Sunday even if the day is perfect. It demonstrates their trust in the Lord to wait another day.

THE CLERGY

The clergy we spoke with in farm communities were keen observers of local habits. They had to be. It was essential to their ministries to know how local funerals were to be conducted and to understand when farmers would come to church meetings and when they would not. Many of the clergy we met had grown up in rural areas or served in rural churches for a long time. Clergy have the added advantage, though, of holding leadership positions and having ties outside the community with other clergy and church leaders.[5]

The clergy were generally pleased to be serving their communities. Despite the fact that small-town and country churches were lower on the pecking order than newer, better-funded congregations in cities and suburbs, the smaller churches remained attractive for a number of reasons. Some of the clergy had grown up on farms and enjoyed living in farming communities. Others were married to farmers or had spouses with jobs in the area. Still others were nearing retirement or were semiretired and preferred a smaller congregation with fewer activities.[6]

None of the clergy were serving congregations composed 100 percent of farmers. Those days were long gone. The churches that were still located in the country included members who lived in town. Their memberships were small enough that the pastor was likely to be serving more than one congregation. Most of the churches were located in towns and included retirees, shop owners, and a few newcomers. The newcomers were the occasional older couple who were rediscovering their rural roots and the less than occasional family that was down on its luck and searching for inexpensive housing.[7]

The implicit comparisons the clergy drew between farmers and townspeople in their congregations nearly always put farmers in the best light. They described farmers as hard-working, conservative, loyal family members who tried to be good stewards of the land. They viewed farmers as the families who had been in the community for generations and provided it with stability and strong traditions. Farmers had deep roots in the land, they said.

"They are good people," was how one of the pastors described the farmers in her congregation. "They have a sense of being here for each other—that neighborly thing you have on the farm. They're second and third generation. They've got stability about them. They're rooted."

Being there for each other was something the clergy knew about firsthand. The priest at a parish in a predominantly Catholic corn-belt community gave a typical example. A farmer in the parish was in the hospital having surgery. "I came to church," the priest recalled, "and said, 'His corn needs harvesting. What do we do about it?' One person stood up and said, 'Let's harvest it.' Everybody came with their combines and trucks and harvested it."

The incident, the priest thought, demonstrated something special about how the Holy Spirit worked in the community. "In the city," he conjectured, "people would be dying and nobody would care because you don't want to poke your nose into something that will cause you trouble. But here, everybody presumes it's their problem."

The clergy also recognized that farming was changing. There were fewer farmers in their congregations than there had been a generation ago. The farmers still in the community were farming more land and using larger and more expensive equipment. The congregation's rural members included more families who lived on small plots outside town but were not engaged in farming.

These changes affected the social dynamics of the congregation. Members were less likely to be interrelated than in the past. Sometimes that was refreshing. Members' work schedules more of the time included off-farm jobs and both spouses working. At the same time the farmers and farm traditions still seemed to set the tone in congregations. That meant greeting one another during the week ("you do the farmer wave," one pastor explained), getting along, supporting the congregation financially, and for the most part minding their own business.

If social relationships were changing, several of the clergy entertained the possibility that farmers' theological convictions might be shifting as well.

The idea was that farmers in the past had been more aware of their dependence on God than they were now. The reason, in this line of thinking, was that farmers were perhaps more in control of their own destinies than they had been in the past. To be sure, they were still at the mercy of Mother Nature. But maybe they felt more in charge as a result of bigger machinery and better technology.

Another priest who served a parish in the corn belt articulated this idea. He had been raised on a farm. He remembered his father standing on the porch watching a bad thunderstorm coming in from the west and worrying that the year's crop would be hailed out. He thought people in those days prayed to the Lord to let something like that pass. "As a young kid," he said, "you were raised with the mentality that you knew you weren't in charge." He was unsure, but he thought maybe all that was changing. If so, it did not bode well for the future of the parish.

The key frustration for most clergy was the same one that many of the farmers identified. It was difficult to get people to attend regularly, let alone take an active part in other programs. Whereas the farmers identified this mainly as a problem to be lamented, the clergy felt more responsible for doing something about it. They, too, attributed the problem to competing interests, such as ball games and travel, and they recognized that farmwork often took priority. But they thought surely something could be done to make the church more central.

One of the pastors whose congregation was a mixture of farmers and townspeople tried canceling the Sunday morning classes that had been held before or after the worship service for as long as anyone could remember. It seemed to him that people were willing to come for an hour but were anxious to get back to work or spend the day with their families. He shifted the classes to Wednesday evenings. That basically failed as well. The farmers especially did not want to drive into town in the evening after long days at work.

In one of the truck farming communities we talked with a Latino priest who was equally frustrated. The parish had taken account of the fact that a growing segment of the population who worked on the truck farms was Latino. He had been hired several years ago and was currently conducting several Spanish-language masses each week. But he said the response was disappointing. During the busiest months the workers were required to work seven days a week. And during the off-season many of them returned to Mexico or Central America.

A theme stressed by nearly all the clergy we spoke with was that somehow the worship services and other church activities needed to be reimagined in ways that would attract young people. They realized that the demographic trends were against them. There simply were not as many young families with children as there were older people. Still, they considered it important to make church as interesting as possible for young people. They knew the future of the church depended on it.

The problem was that the innovations they thought would attract young people butted against the traditions that had been in place for generations. They knew this. They recognized the importance of upholding and respecting the congregation's traditions. But they considered it their place as leaders to challenge the traditions. Usually that meant bringing in ideas they had heard about in seminary or had experienced in nonrural settings.

A good example was one of the mainline Protestant pastors we spoke with who lamented the fact that people in his congregation were too quiet about their faith. He understood that farmers preferred not to wear their religion on their sleeves. He knew this, but he was convinced on theological grounds that people needed to be more expressive about their faith. How could the congregation truly be a faith community if people were unwilling to share their concerns with one another, he wondered.

His idea was to open up the worship service. Encourage more voices. Give people an opportunity to share their testimonies, tell their stories, and voice their prayer requests. He especially hoped the plan would be appealing to young people. It did win over a few of the more talkative congregants. But for the most part it was destined to fail from the start.

Another pastor had only modestly greater success at altering traditions. His congregation of a hundred members was composed of farm families whose ancestors had come from Norway in the 1890s. The Sunday morning services of preaching, hymns, and the Lord's Supper were so familiar that the pastor despaired of initiating any changes at all. When he asked the congregation, though, the congregation told him to give it a try. They wanted to cooperate with whatever he suggested.

Like the other pastor, he too thought it important for church people to be more expressive about their faith. It was not good enough just to sit and listen to a sermon each week and mumble the creeds together. Even though the congregation was part of a mainline denomination, he thought it should embrace the enthusiasm he saw in evangelical and Pentecostal congregations. He challenged the congregation to start raising their hands in praise when they sang. He found it encouraging that some of them did.

A few of the pastors we spoke with thought it important to make religion more vocal in other ways. They were especially concerned about abortion, pornography, homosexuality, and what they regarded as secularism in the public schools and government. Hardly any of the farmers we spoke with were actively involved in these issues. But the clergy who were involved felt it was their responsibility to speak up. They did so from the pulpit and occasionally at town council and school board meetings.

Notwithstanding the various tension points, the relationships of clergy to their congregations were by most accounts congenial. As community leaders, the clergy were in charge of organizing soup kitchens, visits to the elderly, and special programs on Christmas and Easter. They conducted funerals, performed baptisms, and were on hand to counsel the bereaved.

The clergy who had served in rural communities for any length of time understood the special challenges of ministering in places where population was sparse or declining, members were older, and traditions were strong. They were less sure that church leaders elsewhere understood these challenges.

It worried them that foreign clergy with little experience with local customs were increasingly staffing rural churches. It was easy to chuckle about conversations at denominational meetings with urban pastors who had no idea what an acre was or why a combine had a header. They found it more disturbing to learn of denominational plans to shut down rural churches.

When asked what they liked about their clergy, farmers' most common reply was that their pastor understood them and cared about their families. They appreciated a good sermon and they especially valued sermons and lessons that were upbeat. It felt good when crops were bad to hear about hope and to be reminded of God's love. It would have been surprising if farmers had mentioned anything else.

They did mention something else, though. The pastor was frequently their connection with the outside world. Not that farmers were particularly isolated. They usually had relatives living in other places and working in different occupations. And like everyone else, farmers traveled and took vacations. But clergy had different connections. They knew about missionaries serving in other countries. They went on international mission trips. They talked of working at inner-city churches and of experiences during seminary. It was refreshing to hear about these experiences. It was especially reassuring when the community was small, when the local population was declining and more of the pews on Sunday morning were empty, to have a pastor who was part of a larger picture.

As is true of congregations in nonfarming communities, the most meaningful parts of congregational life in farming communities, though, were local. The activities that mattered were the worship services that brought people together once in a while, the potluck dinners and work days that made the church building seem like a home, and the small talk before and after church meetings that reinforced social ties. All of that necessitated paying close and respectful attention to the rhythms of farm life.

A seasoned pastor at one of the churches we visited expressed this understanding particularly well. "I would say there are two things that drive the families here," he said. "It would be farming number one and church number two." He wished it was the other way around, but he understood that farming had to come first. "As a pastor, you have to schedule things, events other than worship, you have to schedule it around what's going on at the farms."

"They'll be out in the fields till past dark. They'll come to church, but you really have to be mindful of the people's lifestyle." As one of the members told him, "If you want the church to survive, I have to have my farm running." The pastor knew this was true.

MISGIVINGS ABOUT RELIGION

National polls suggest that Americans are becoming less religious. At least the fraction who do not identify with any particular religion is larger than it was. In some polls as many as a fifth of the public gives this response. What to make of these polls, though, is hard to determine. Americans are still more religiously involved than people in most other countries. Many of the people who disavow being affiliated with a particular religious tradition still assert that they believe in God.

For the reasons mentioned thus far farmers would seem disinclined to reject religion. At least many of the ones we spoke with felt this way. Some were quite sure religion was particularly important; others thought polls conducted in their community would resemble national averages. They agreed that religious beliefs and practices were important, even though the demands of farming sometimes got in the way.

That makes it interesting to hear farmers express misgivings about religion. In candid remarks they acknowledge frustration. They note the standard complaints that people of faith have always made. The preaching is far from inspiring. The pastor should have pursued a different line of work. The priest should have moved on years ago.

"All my life I've been involved," a farmer whose feedlot and pastureland are fifteen miles from the Baptist Church he and his wife belong to says. But the "rookie preachers" the church has hired in recent years have left him cold. "I can be mad and raise Cain and try to get them fired," he says. Instead, he has decided, "I'm just not going to go."

"Do I have a church," he asks. "Yes, I have a church. I'm not going to give up on it. It's like a problem child. You love them forever, but you're just kind of exasperated with them." For the time being he is feeding his spirituality in other ways. "There are some Christian radio stations around," he says.

Misgivings also stem from conflicts in congregations and from disappointments bred of congregational politics. These are hard to get over when congregations are small, declining, and poorly staffed. A disgruntled suburban member could simply shift to a different congregation. Nobody would know. In a farming community, everyone would know.

One of the farmers we spoke with was still piqued about a criticism the local church people had made about his brother for not attending regularly or believing as he should have. This had happened years ago. The man telling the story still went to church but made it clear that he would not serve on any committees.

"The leadership changed," another farmer who had been active on committees at his church for years explained. It wasn't as comfortable being at the church as it used to be. "I worked at it a while," he said. But it wasn't working. "I chose just to kind of drop out." He and his wife now attend at a different church in a town ten miles away.

Yet another farmer said things were awkward at the church he and his wife attended because it was composed of several large families who thought they were "higher and mightier" than everyone else. He felt guilty mentioning this concern and would not have done so except that the interview was anonymous. It seemed petty but was still important.

"I know it's what your relationship with God is," he said, "but it's just kind of hard to go to church and worship when people sitting all around you are kind of. . . ." He wasn't sure how to put it. "What I'm trying to say is we have been very poor church attenders the last few years. Right or wrong."

He says his belief in God and Jesus has not changed. It sustains him especially at times when there have been farm accidents or family illnesses. "I guess we find our faith within ourselves," he says. He does not mean his faith is *in* himself, alone, or that it lacks a relationship with God. He means that faith no longer depends on being active in a congregation as much as it did earlier in his life.

When God is taken for granted, as it is in some of these examples, it can be disconcerting to think that time actually has to be spent worshipping and praying and helping staff church committees. It is hard enough to make time when farming itself seems to take up the hours. Harder still when the crops fail and the cattle die.

The Loeschers were one of the couples we talked with who were less than sanguine about religion. They may have had more reason to complain than most. Not only was the dairy business in their community struggling. The land they rented could be taken away. Some of it was owned by a preacher. They trusted the preacher. But it bothered them when he went around saying "bless this" and "bless that." They figured he meant well, but when things went wrong, they felt like saying, "How come God didn't bless me?"

Nothing in these remarks suggests that farmers are abandoning the religious traditions that have been so important to farming communities in the past. Like changes in families and among neighbors, though, rural churches are being affected by the shifting tides of rural communities. Faith may still be part of the landscape, but there is little reason to think that farmers are particularly devout or that farming somehow lends itself to superstition about the role of incantations in making crops grow and animals fertile.

Faith of this kind is neither inimical to modern farming methods nor particularly conducive to them. Students of agrarian life in earlier times saw a connection between superstition and a limited capacity to learn how to farm in modern ways. The early modern linkage of ascetic Protestantism to rational acquisitive capitalism could be said to have propelled progressive agriculture just as it did commerce and industry. Yet the same argument suggested that religious motives would cease to be as important as capitalism advanced.

The farmers we spoke with mentioned no connections of farming to the anxieties about predestination that arguably motivated an earlier generation of entrepreneurs. In current thinking God appears neither to reward the just or punish the wicked but to provide spiritual strength. The intellectual underpinning of this strength is not a repertoire of divine explanations for abundant yields or poor crops. It is rather an abiding background conviction that things happen for a reason, even though that reason remains unknown.

Farmers vary, just as other Americans do, in how and how often they pray and in what exactly they believe about God. Over the decades the dominant faiths founded congregations by the thousands in rural communities, and many of those congregations remain, albeit in diminished versions. Newer groups with Pentecostal and fundamentalist leanings have gained ground in towns and suburbs adjacent to farming communities as well.

Participation in these congregations is shaped by practical considerations as much as by personal piety. Farming in close proximity to parents and grandparents encourages multigenerational loyalty to congregations as well. The same loyalties can be fractured by family conflicts. Putting the loyalties into practice depends on time and energy that many farmers say is in short supply.

They may not go to church as regularly as their parents or grandparents did. That bothers them, especially when they grew up with stories and visual reminders of ancestors who were indeed devout. Believing that a person's inner spirituality matters—that what is truly important is one's relationship to God—helps to shore up the conviction that a person is still in good standing with God.

It helps to say that religion is less important now than it was in the past because of changes in ethnic traditions. Those traditions reflected an earlier time, the argument goes, when immigrant customs and extended kin networks were more important. A person can still have a relationship with God in other ways.

Being able to laugh at yourself assists as well. A woman in her sixties gave an interesting example. She was explaining somewhat sheepishly to her eight-year-old granddaughter that she and her husband had not been going to church much lately and felt rather guilty about it. "Oh, grandma," the girl replied, "are you a churchaholic?"

The more important arguments about going or not going to church reflected farmers' view that God was a constant element in their lives, no matter what they did or did not do. Rain would or would not come whether they prayed or went to church every Sunday or seldom went. The point was not to bargain with God but to be responsible in carrying out one's work.

"The ox is in the ditch," a wheat-belt farmer explained. "It says that right in the Bible." He felt guilty attending church services as infrequently as he does. But he takes solace in Jesus' remark that rescuing an animal from harm is sufficient reason to violate the Sabbath. "Is the ox in the ditch every damn Sunday?" No, he says. But when the fence is down, the calves are out, and the fields are blowing, it often seems that way.

INDEPENDENCE

4

We're independent folks. We're the last real cowboys, so to speak. The cowboy riding across the open range. Nobody takes his freedom away. The farming community is almost that way.

—Cotton-belt farmer, male, age 65

I like to be able to do what I want to do when I want to do it. That's one of the big draws to farming.

—Wheat-belt farmer, male, age 70

"I guess I just like being my own boss." This was one of the most common answers when farmers were asked why they had chosen to farm and what they still like about farming. But it is impossible not to wonder what farmers mean when they say they are their own bosses. Is this merely a cliché? Are they discounting the extent to which their activities are dictated by landlords, government regulations, and the weather?

Asking about farmers' sense of being independent is an opportunity to consider one of the oldest features of American culture. From earliest years to the present, observers of the United States have argued that American culture is individualistic. Americans take pride in having freedom as individuals to do as they please. And with individual freedom comes personal responsibility. At the same time individualism poses concerns. Critics worry that American individualism undermines civic-minded orientations favoring the common good.[1]

When these aspects of American individualism were first identified America was predominantly a nation of farmers. Farmers valued having their own land and being self-sufficient on that land. They may have felt strong obligations to their families and they may have extended the milk of human kindness to their neighbors. But they seemed to be especially interested in upholding their individual autonomy and taking responsibility for themselves.

Over the years observers of American culture have identified several prominent and sometimes conflicting strands within this larger ethos of individualism. One is a kind of self-interested strand in which the pursuit of individual success takes precedence over all other commitments. Another consists of strong-willed inner conviction about what is right and about what a person must do to pursue that which is right and to avoid doing wrong. In recent decades different strands have been identified. One is best termed simply as narcissism—an obsession with one's self and with deciding amid cultural uncertainties who one is and who one wants to be. Another is a kind of expressive individualism evidenced by interests in paying attention to one's feelings.[2]

Farming is hardly the place to find some of these variants of American individualism. It is an unlikely location for the kind of shallow self-interested pursuit of individual pleasure documented in some studies or of the therapeutic motif described in others. The sense that farming is particularly attractive because of possibilities for being one's own boss, though, is worthy of special consideration. If that kind of independence is eroding, farming as an occupation may be less attractive than it was in the past or farmers may be in the process of inventing new arguments about why farming is still appealing.

Research in other contexts has approached questions about individualism and personal independence in several distinct ways. In the 1930s a popular way of distinguishing people who were more independent or less independent was to say that the former were introverts and the latter were extroverts. By the 1950s the idea of innate personality types gave way to arguments about the effects of early childhood development and socialization. Americans were said to be highly individualistic because of parents training their children to be independent.

Other approaches have emphasized the presence or absence of social networks. In this view personal independence was associated with a lack of such networks, while the more preferred style of interaction involved extensive social relationships, which not only provided support but also entailed responsibilities and encouraged altruistic behavior.

The weakness of those approaches was that they slotted people into fixed categories. Because of biological or psychological or sociological characteristics, people were classified as relatively more or less individualistic. The more recent approaches, in contrast, have emphasized the complexity and malleability of personal independence. In this view individuals play a more active role in constructing their own understandings of themselves.

The constructivist approach recognizes that life experiences are sufficiently varied that they can be interpreted in multiple ways. These interpretations are significantly composed of accounts. Accounts are narratives that explain how and why a person came to be a certain way, whether the question at hand is the choice of a career or marital partner or some other decision such as where to live or what to wear. Plausible accounts explain the connections among a person's circumstances and choices.

We listened closely as farmers talked about their values and their daily lives to see how perceptions of independence came into their narratives. The point of entry for understanding these perceptions is how they chose to become farmers in the first place. Being one's own boss makes sense only when the most relevant occupational choices are considered and when various other options are ruled out. Then on a day-to-day basis farmers' emphasis on independence affects the decisions they make, the risks they are willing to take, the scale of their operations, their investments in machinery, and how they interpret success and failure. Beyond that, farmers' convictions about personal independence color their interpretations of the wider social landscape as well. All of these understandings are currently in flux, shifting as farming itself requires new skills and is increasingly governed by technological innovations.

CHOOSING TO FARM

The hallmark of being an independent person is being able to choose. We value the freedom we have as adults to make our own choices about how we spend our time and money. We encourage children to grow up thinking for themselves and making their own decisions. One of the most significant decisions people make is choosing their occupation.

Farming is interesting in this respect. So many farmers are following in their parents' footsteps that the question arises of how independent their choice to farm actually was. Did they feel they had to farm because their parents wanted them to farm? Or did they feel they were truly making a choice?

Farmers' accounts of becoming farmers certainly acknowledge the constraints shaping their decisions. Many of the constraints were economic. It had not been possible to attend college or to graduate or pursue a career requiring an advanced degree. One of the farmers we spoke with, for example, had aspirations to become a veterinarian but simply did not have enough money to stay in college. He decided to farm with his father instead, hoping to gradually expand enough to get married and start a family. Others chose to farm because the alternative was working at a low-paying manual-labor job.

Family ties presented another constraint. Having lived in one community all their lives, they wanted to stay there. They felt loyalties to parents and siblings. Some stayed because a parent died or was too ill to continue farming.

It was possible nevertheless for nearly everyone we spoke with to say that they had *chosen* to farm. It was a decision they had made, sometimes more thoughtfully than others, but always with the knowledge that they could have taken another path. The choice was one of the most important decisions they had made. They understood it as a reflection of who they had wanted to be when they were young and of who they already were.

Decisions like this are complex enough and they usually happen over a period of years that begins in childhood and extends into early adulthood. That makes it difficult to sort out the relevant factors. The significant aspect, though, is how people in retrospect interpret their decision. Do they develop a narrative that emphasizes their independence in deciding how to spend their working life? Or do they focus on the circumstances that shaped their decisions?

The warm memories farmers recount of growing up around farm animals and driving the tractor or learning skills by helping with the chores suggest that narratives about choosing to farm are filled with rosy notions that farming was simply the ideal choice. There was some of that. But farmers who grew up on farms knew the negative aspects of that work as well.

"I didn't like the hogs!" a corn-belt farmer exclaimed. "They're supposed to be a clean animal. They don't let it go anywhere like cattle. They'll go dump in the corner. But they still get pretty stinky. And cleaning the chicken house, that was the worst!"

A wheat-belt farmer recalled milking five cows by hand before school when he was a first grader. In fifth grade he started raising pigs. It was his way of earning some spending money. He realized eventually that he loved animals and wanted to farm. But at the time it was difficult. Everyone else at school got to be in sports or go to the games. He did not. He knew that farming would be hard work. The likelihood of earning much money was small.

Choosing to farm was thus an exercise, as farmers look back on it, in rational decision-making. It involved weighing what was good about it against what was not so good.

Weighing what was good about farming against what was not so good also meant considering career options other than farming. The fact that so many farmers grew up on farms and now farm land their parents or grandparents owned does not mean that they considered no other lines of work. Usually they had the firsthand experience of siblings making choices other than farming. They often worked part-time at other jobs.

Like many decisions in life, choosing to farm made sense for the farmers we spoke with not only because it offered the most rewards and came with the fewest drawbacks but also because it seemed most authentic or natural. It was the choice that seemed right in terms of who they were as persons.[3]

"There's never been an alternative career path," a fourth-generation wheat-belt farmer in his early forties says. He remembers the exact day as a freshman in high school when the teacher asked the class to write down what they wanted to be when they grew up. Some of his classmates wanted to be doctors or lawyers or teachers. "I wrote down that I was going to be a farmer."

He says this with no regret. Lacking an alternative career path apparently did not seem limiting at the time. Three decades later it still does not. He is proud to have known all along that he wanted to farm. Being a farmer is consistent with his understanding of who he really is.

There are many ways of making sense of one's decisions in terms of personal authenticity. Personality tests and personal interest surveys are one such method. High school and college students frequently take such tests and surveys to learn if they have personalities and interests that feature interpersonal skills, mechanical skills, and the like.

These tests are meant to give guidance at the time. They assess whether a high school student has a preference for one kind of occupation or another. But the tests also serve a retrospective function. We talked to farmers in their fifties and sixties who still used the tests they had taken in high school as evidence that farming was best suited to who they were as persons. The test results legitimated their choice.

Another perhaps more common sense–making activity is to formulate narratives that explain in essence that a choice seemed right because it reflected "who I really am—the real me." Studies of religious converts, for example, show that converts often explain their decision to switch religions on grounds that the new faith puts them in touch with who they really are. The farmers' version typically connects their decision to farm with happy childhood memories and with coming early to the thought of wanting to farm.

This sense of farming as the choice most consistent with "who I really am" was one of the most common narratives among the farmers we interviewed. It came closest to illuminating the meaning of farming being "in the blood." Farming in this understanding is a choice driven by how and where one was raised, what one's parents did for a living, and the options available or not available. It is in this respect not much of a choice at all. It is determined by all the contingencies of one's upbringing. At the same time it is understood as a choice, and indeed as the right choice, because it reflects who one is and thus enables that person to do the things in life that are most natural.

Mr. Engstrom, the fifth-generation farmer we met in chapter 2, illustrates this kind of narrative. When asked why he got into farming, he says, "There never was a question of doing anything else." He does not mean he had no choice. He attributes his choice not to his family history, as he might have, but to a teacher in high school who taught vocational agriculture classes. "My ag teacher was really good at reading what each boy needed," Mr. Engstrom says. "He told us, 'Now, I know all you guys aren't going to be farmers.' Out of my ag class, one of them ended up being a high school principal and another ended up being a teacher. Only two of the fifteen became farmers."

The point of the story is that farming was the choice that best suited who Mr. Engstrom was. "Whatever you were good at," he says, the ag teacher "would head you in that direction and work with you in that way. If you were good at something else, he would take you in that direction."

This is an example that illustrates the special power of role models who also serve as mentors. Among the farmers we spoke with whose parents had farmed, the parents typically served both these functions. The child emulated what the parents did and gauged whether the activities and skills involved felt right. The parents' mentoring function included not only teaching the child those skills but also talking over the dinner table about what was good or not so good about farming.

A slightly different way of showing that the decision to farm was almost a given and at the same time involved choice is evident in Mr. Rayburn's account of how he became a cotton-belt farmer. When asked his reasons for going into farming, he begins with the assertion that "it's in your blood." He elaborates, explaining that the decision reflects "the way you were brought up and the way you were taught." Were that all, the implication would be that he had little choice in the matter. However, farming being in one's blood and being rooted in one's upbringing also means that it makes sense for him. It is consistent with who he is.

As further evidence that farming reflects who he is, Mr. Rayburn adds, "I like the opportunity that I'm technically my own boss." This is his way of saying that farming provides a good dose of personal independence. "Technically" qualifies the amount of independence he feels. He says he has a lot of obligations and mentions specifically a full-time employee he feels responsible for and supervises. The employee, though, has been around long enough and grew up in the area, so the amount of supervision needed is minimal.

Independence for Mr. Rayburn ultimately means being able to do different things from day to day, facing different challenges, and enjoying those challenges. "It's not just getting on the tractor and planting," he says. "Each morning you'll wake up with different headaches or different successes and different rewards. You don't know what is lurking around the corner, but there's rewards as well as hardships seems like every day."

This is an interesting and important way of understanding what it means to be one's own boss. Both Mr. Rayburn and Mr. Engstrom emphasize that a person feels most free, most independent, most fulfilled when doing something enjoyable, however difficult it may be, because it fits with who one basically is. The idea is simply how a person might describe a garment that fits. If it fits, it wears easily. It does not feel uncomfortable.

Independence in this sense is not rugged individualism. It does not imply going it alone, doing whatever one damn well pleases, come hell or high water. Being independent does not come at the expense of defaulting on one's obligations. The personal freedom implied in being one's own boss is compromised only when social obligations force one to do something inconsistent with this understanding of one's true self.

There was another sense of independence, though. For many of the farmers we spoke with, farming carried the added attraction of limiting the extent to which they had to interact with other people. To be sure, they varied in this regard, just as people in other occupations do. But when asked if they were the kind of person who liked to be around people or alone, the majority said they truly preferred to be alone.

Even the ones who were involved in business dealings that required daily interaction with people or who observed that one of the joys of farming was being in constant contact with family members said they felt most at ease working alone all day on the tractor or enjoyed feeding the livestock alone each morning and evening. Others acknowledged that they liked to sneak off in the evenings during slack seasons and tinker with machinery in their shop.

"When you're on the farm, you're just there talking to the tractor, talking to yourself, doing a lot of thinking," was how one farmer put it. He likes that, being alone, having time to think. He acknowledges, though, that sometimes the conversations with himself need a break. Then he goes to town, sees people he knows at the restaurant, and appreciates the chance to hear what's going on.

The sense of being alone and enjoying it is quite different on farms than it might be for a city person who strategically keeps people at bay. It differs from the notion of being in a lonely crowd with people all around and wanting relationships but still feeling alone. Farming, as farmers see it, is a legitimate way of being apart from other people.

Mr. Bower was one of the most adamant about the connection of farming with liking to be alone. "There is no such thing as a farmer who is an extrovert," he observed. He had planned early in life to become a schoolteacher or college professor and he had worked in a retail store for a year or two after college. But when the opportunity to farm came his way, he jumped at the chance.

At a bare dry-land wheat farm with few trees or farmsteads anywhere in view a farmer told us that he was five miles from the nearest town but "would like to live farther out." He said his nearest neighbor was a mile away, adding, "I think it's a mile too close." He goes into town often enough on business, but his desire to be on his own runs deep and is not only possible but also reasonable by virtue of living in the country. "We love agriculture," he says, "but it's the lifestyle we really enjoy."

DAY-TO-DAY ACTIVITIES

The farmers we spoke with emphasized that the day-to-day activities involved in farming truly enabled them to be independent. They had well-used idioms and scripts to draw from, rhetorical devices that provided the connective tissue between what they did and how they interpreted what they did. The part of farming they liked best, they said, was being in control, taking charge, being their own masters, counting on themselves. Being one's "own boss" was more than a pat answer to why farming was a desirable occupation. Farmers could point to the tasks of a typical day as evidence. These activities required them to be independent. They did so in three ways.

First, farming was an independent activity because the farmer worked alone most of the time and was free to decide how to spend the day on various activities. Second, farming involved making major decisions from day to day, rather than having someone else make those decisions. And third, farming involved such a diverse set of skills that it perhaps uniquely demonstrated how a person could tackle many things independently rather than having to rely on others.

"Basically the independence," a cotton grower who started farming with his father in the 1970s replied when asked why he became a farmer. "I'm kind of a solitary sort of person and farming's a fairly solitary occupation." That was certainly believable. Cotton fields stretched across the flat reddish soil around his farmstead for as far as the eye could see.

Working alone is good if a person does not enjoy being with people or prefers not having people around most of the time. As one farmer put it, "You are removed from somebody bugging you. It gives you space." It also provides testimony to one's independence. Consider the farmer who spends forty-five minutes every morning feeding cattle or the farmer who sits on a tractor for twelve hours planting corn. Those tasks have to be done. But nobody besides the farmer says they do or dictates exactly when.

Mr. Loescher's love of dairy farming is that he likes being his own boss. In practical terms that means working alone except for the company of Mrs. Loescher. He likes being with the cows and working outdoors. "There is never anybody telling me that I have to do something," he says. He can clean the mangers really well one day and then less well another day if he decides something else is more important.

The cotton grower elaborates on why he appreciates farming being solitary. "I'd rather work by myself than among a bunch of people," he says, "especially if it was a team where you had to play a part and everybody was comfortable with that." As a farmer, he adds, "You have your own deal. You're the guy."

Having your "own deal" is appealing because it signals a person's place in the local hierarchy. Not everyone can be "the guy." Those who can are people who have made it. They have been able to maintain their freedom.

Those who cannot be the guy must work for someone else. "My dad had these two classes of people," he continues. "There were people in business for themselves. They owned the wages. And there were those who worked for wages. They were in second place."

That sense of who was on top and who was in second place, he says, was instilled in him growing up. The wageworkers were in many respects admirable. They were "fightin' men." But he also learned from his father to look down on them. There was "kind of a scruff image" about them. "So as long as I was working for somebody," he recalled, "I felt frustrated."

Being a "one-man operation," as farmers say, not only gives a person flexibility to start the day fifteen minutes late or to cultivate one field first instead of another. It makes sure things get done right and on time. The farmers we talked with insisted that owner-operators were the ones who cared the most about doing good work. Hired workers were not. As one farmer acknowledged, "I have a young hired man and there are days I'd just like to drown him!"

"I'd be an awfully poor employee if I had to go punching a time clock for somebody," another farmer acknowledged. He spends a lot of time in his "man shed," even sleeping there some nights, comfortably alone with his tractor and farm tools. He says the highlight of being his own boss is the daily freedom it provides. "I'm free to go start in the field at eight o'clock in the morning or eight o'clock at night as long as I get it done. There's nobody telling me when, where, and what." He began farming when he was thirteen, renting some "crap ground" from an uncle, losing his crops and hogs on more than one occasion, but never regretting the life he has chosen.

A farm couple in their fifties we talked with who grew wheat and tended livestock got to laughing about what it meant to be independent and why that was important. "I don't want somebody telling me when to do this and when to do that," the man declared. "If I want to go out and work past dark to get stuff done so I can take off the next day and do nothing, then that's what I'm going to do!"

His wife agreed it was nice not having someone tell you "these are the six things I expect you to get done today." It sounded almost too good to be true, which made her chuckle. The more serious aspect of farming, she thought, was that you knew what had to be done. Being your own boss meant having the requisite knowledge and experience. "You just get up and go do what needs to be done," she said. "Everybody working at the farm basically knows what needs to be done, so everybody can just go do their own thing."

The point of emphasizing flexibility in their daily schedules was not to suggest that farmers could linger over morning coffee or go fishing whenever they want. It is rather that decisions have to be made. And an independent person takes responsibility for making those decisions. Nobody else is there to make them or to see that the work gets done.

In this respect farmers are independent not only because they have choices to make but also because success as an independent farmer necessitates making good choices. Some of the farmers we spoke with emphasized this point by contrasting themselves with office workers working at nine-to-five jobs. A person like that, farmers said, merely put in the required hours and went home. In contrast, farmers would work until midnight if it meant getting the corn planted or the cotton harvested.

Major decisions demonstrate independence even more clearly. Many of these decisions are made in consultation with loan officers, landlords, equipment dealers, agronomists, and relatives. Farmers, though, are like other business owners and executives. When it comes to making important decisions, they are the person deciding. Statements such as "I built a new shop" or "I bought a new combine" or "I rented a piece of ground" are more common than statements about consultations and collective deliberations.

"I worked at a company that made pizza toppings," a woman who farms with her husband recalls, "and there were people there who ran a slicing machine eight hours a day. All they did was slice pepperoni. That would just be so boring." She appreciates the variety farming provides. "You may do a really hot, dirty job, but you only do it a few days a year or a few hours a day and then you go do something else."

Her main point, though, is different. She did not work as a pepperoni slicer. With a college degree in accounting, she handled the company's finances, taxes, and public stock offerings. It was a large company, and she held a responsible management job that included plenty of variety. She earns less hour-for-hour in farming but prefers it because she is her own boss.

She says she and her husband spend long hours making important management decisions. She mentions major decisions involving the purchase of a new grain drill and estimating the relative costs of no-till farming. The decisions are mentally challenging. They are also rewarding. She feels invested in them personally. She gets to see them from beginning to end. The full story.

A corn-belt farmer elaborated on a similar idea. His spread is large enough that some of his farm improvement projects cost hundreds of thousands of dollars. This is the part of being his own boss that he especially likes. "I really enjoy the organizing and planning and seeing it develop," he says. A recent example was tearing out some older grain bins and putting in a new farm shop. He did all the excavating with his own equipment. Then he had a cement foundation poured, got an electrician to put in a heating system to provide floor heat, and lined up another guy to put up the building. "Just seeing a project develop like that and overseeing it," he says, was enjoyable. "It's kind of fun when things fall in place. It's a sense of accomplishment."

Whether the decisions were small or large, the sense was that farmers' freedom to make day-to-day decisions was greater than it would have been in any of the alternative lines of work they could seriously imagine. Those imagined alternatives were different among farmers with different levels of education. For those with no college training, the possibility of having to work in a blue-collar occupation implied literally having a boss calling the shots. Compared to farming, working in construction or at an assembly plant suggested little freedom to make one's own decisions.

The farmers who had gone to college were more likely to draw comparisons with workers in managerial positions and the professions. The contrast was reminiscent of portraits drawn in the 1950s of an emerging American culture dominated by "organization men" and "other-directed" personalities. It connoted a new kind of boss to which even well-educated workers were subject. The boss was a committee. Decisions had to be made—and indeed were assumed to be made better—in consultation with a committee of peers who brought their specialized perspectives and opinions to the table.

Consider how Mr. Hebner, the wheat-belt farmer we met in chapter 1, describes what he likes about farming. He and his wife are college graduates. When he started college he thought seriously about majoring in architecture and pursuing a career working for an architectural design firm. But he somehow realized even then that he was not that kind of person and now says he is far happier farming. "If you work by yourself," he says, "you know why you do the things you do and you don't share those reasons with a lot of people because you don't have to."

He believes it is important to have good reasons for the decisions he makes. He draws on his college training, reads extensively, goes to farm meetings, talks to his neighbors, and consults with experts about agronomy and business conditions. The issue is not a preference to make decisions in a vacuum but of not having to make them in a committee. "At the end of the day," he says, "you're the one who has to make the decision. And if you've made the decision because of well-thought-out reasons, why discuss that with everybody?"

His reference to "everybody," judging from what else he says about farm decisions, is meant to suggest that a few conversations with trusted interlocutors are advisable, but discussions beyond that are unnecessary and perhaps even counterproductive. At that point they seem excessive, as if the decisions are being explained to "everybody."

This way of thinking about personal independence illustrates a particular view of rationality that came up repeatedly in our interviews. It emphasizes a means-ends variety of rationality in which the rational course of action is determined by collecting all possibly relevant information and then judging the action on the basis of whether it produces the desired results. It contrasts with varieties emphasizing procedural, deliberative, and communicative understandings of rationality. In those versions the rational course of action is the one decided upon collectively by following a specified set of procedures and discussing them with a relevant committee of peers until arriving at an agreement.

For Mr. Hebner the sharpest contrast was between having good reasons and having to discuss those reasons. Although he was fully capable of articulating his reasons for making particular decisions when we asked him to do so in our interview, that was not a skill he considered essential to being a good farmer. Being independent as a farmer meant not having to devise persuasive arguments of the kind that would carry the day in a corporate setting involving committee meetings. He was the kind of person, he felt, who could think through a complex problem and come up with the right decision. He did not view himself as the kind of person who could go out and sell those reasons to other people. Those were the skills of lawyers and politicians and salespeople, not farmers.

Few of the ways in which farmers described their daily activities provided a clearer contrast with other occupations than these descriptions of how important decisions are made. Certainly in university settings as well as in business and government contexts the preferred decision-making mode is to assemble a committee. The logic is that more heads are better than one. Committees not only brainstorm complex problems. They also develop accounts of why a particular decision made sense. Those accounts can be supplied to whoever the relevant audience is through meeting notes and reports. In contrast, farmers made decisions on their own, assembling information from multiple sources and discussing ideas informally with family members, but doing so independently rather than by committee.

Having many skills is yet another way in which farmers describe their understandings of personal independence. Farmers say they are a "jack of all trades." Seldom do they add "and master of none." The range of skills involved is truly remarkable. In many other occupations specialized knowledge, however extensive, limits the variety of skills required. An expert welder working for a pipeline company, for example, would not be expected to make major decisions about the company's investments. Or a construction worker driving a backhoe would not be expected to take the place of an expert welder.

Doing everything for one's self was a matter of pride for many of the farmers we interviewed. Self-reliance was part of the family tradition they had learned from their parents and grandparents. It set them apart from and above people they knew who needed help. His father, one man said, had built a new barn all by himself. He was so independent he even pulled the nails from the old barn and saved them for the new one. This was a trait the man telling the story admired. He liked to fix his own machinery and tend a sick cow without asking anyone for help.[4]

"There is nothing on a combine I can't take apart and put back together," a wheat-belt farmer states with modest pride in his voice. He has his own fully equipped shop where he repairs all his machinery. He says he does all the major overhauls on everything except diesel engines.

Being one's own boss in farming implies being able to do all of the various tasks involved without having to look very often for help. The men we interviewed said they especially enjoyed that. They considered it challenging to shift from one task to another and to face new problems from day to day. They were proud to have built their own shop, spent the day welding a broken cultivator, shifted from raising hogs to fattening cattle, experimented with a new variety of seed, and started a side business during the off season.

This desire for independence from day to day is one of the reasons farmers who might have the opportunity to farm on a larger scale choose not to. And if they do expand, they do so by purchasing a larger tractor that they can operate by themselves rather than taking on a partner or hiring farm laborers. "I don't want to have someone else work on my equipment," Mr. Bower explained. "I don't want to be dependent on anyone else for getting my harvesting done. I own a sprayer now because I got disgusted with needing to have [the co-op] do my spraying and not doing it like I would or when I want it." To underscore the point, he added, "I hate working with people too!"

The more significant gratification from being adept at multiple tasks, though, was the sense of mastery involved. Being an independent person, in this respect, is having accumulated a set of skills or a body of knowledge. It is fulfilling to have gotten better at something over the years. Farming in this way of thinking is a practice. It is learned by practicing, just as playing tennis well or becoming a good lawyer is.[5]

A farmwoman in her sixties who is the principal operator with her husband of a 150-acre farm on which they raise beef cattle provided an interesting illustration of this point. They are first-generation farmers who have had setbacks and probably would not have survived in farming had it not been for income from off-farm jobs and a strong passion to stay in farming. She and her husband especially appreciate farming for the independence it provides. She defines independence as gaining a sense of accomplishment from doing things well.

When she and her husband started farming they were in the dairy business. They took pride in learning how to keep the somatic cell count in the milk low and how to reduce the amount of butterfat. After having to sell their herd on one occasion and losing it when the barn burned on another occasion, they have shifted to beef cattle. It makes her feel good to have mastered the new skills involved in this kind of farming.

She is not like Mr. Bower. She enjoys being around people and is perfectly willing to admit that she calls on the neighbors for advice. Still, she especially relishes the challenge of being able to tackle most of the tasks on her own. That includes everything from mowing hay to fixing fences to keeping the farm accounts in order. Most importantly, she says, it means "thinking like a cow."

"Thinking like a cow" means trying to understand the world from a cow's perspective. When the calves arrive on a truck from another state each spring, she knows what they have been experiencing. They are away from their mothers for the first time. The truck ride has been traumatic. The smells here and the view and taste of the grass are different. She spends some time with each one to spot any indications of sickness from the journey or lack of appetite. She identifies the ones that seem to be leaders. That will be important when she divides the herd into different paddocks.

Later in the season, she knows how to gradually shift from grass to hay without upsetting the cows' digestion. She can usually guess when they are about to run out of salt or why they have disturbed the float in the water tank.

This is part of what she means by independence involving a sense of accomplishment. The knowledge she has accumulated over the years is hers. It resides in her person. It gives her a sense of self-confidence. When new challenges arise, she feels capable of meeting them.[6]

The irony about wanting to be independent is that so many farmers are not truly free of obligations to their parents until their parents die. Unlike the young person who goes off to college and then takes a job in a different state, farmers are far more likely to be living just down the road from their parents, farming land their parents own, and having their parents offer advice about how farming should be done.

The tension between valuing independence and being tied to parents is one of the sources of family conflict that we discussed in chapter 1. A farmer in his forties or fifties like Neil Jorgensen who feels that his father still wants to control major and minor farm decisions is likely to chafe at those relationships all the more by virtue of being in a line of work that champions individual freedom.

The solutions farmers we spoke with had discovered varied. One solution involved mentally exempting family relationships from thoughts about independence. Being free meant making decisions without regard to anyone besides one's parents. A second solution was to develop a business model that kept one's parents out of the picture to the greatest extent possible. Many of the farmers we talked with about this emphasized the psychological as well as the economic value of owning some of their own land.

Several farmers said they chose to farm in a different county from their parents. They thought it was good to be close enough to visit but far enough away that they could develop their own reputation and not be viewed as someone's son or daughter. The other solution was to work alone. They might live down the road from their parents and farm family land, but tending cattle alone and spending long days alone on the tractor reinforced their sense of personal independence.

The clearest contrast with this ideal of truly working alone is large-scale farming with multiple family members and employees. At Granger Farms, for example, Tom Granger spends precious little of his time alone. A typical day involves getting the thirty employees started on their various tasks. "Chasing people all day" is how he describes it. He is not even his own boss, strictly speaking. The family corporation is.

For managers of large farms like Mr. Granger, independence has come to mean something quite different from the way small-scale solo operators describe it. Independence means running a business instead of being an employee. He figures he could have used his business skills working for a large brokerage firm in the city. During college he thought about it. But that was an alien world. It seemed confining, almost as if he needed to work outdoors to feel free. He enjoys bringing in a good crop of peas or potatoes or cabbage. That is the part he likes best. The hardest part was working with people.

Actually disliking working with people was not how most of the farmers we spoke with put it, although some clearly felt that way. Their preference was simply to work alone or not have to supervise someone else. Indeed, an advantage of long hours doing fieldwork was that it could usually be done alone. That was the enjoyable part, a wheat-belt farmer explained, even though he routinely worked from five in the morning to nine at night for weeks on end during harvest in the summer and planting in the fall.

There were farmers too who said they enjoyed being with people or at least did not mind working with them. The bottom line, if there was one, was that they felt comfortable working alone some or even much of the time. As one farmer put it, "I enjoy the company of others, but I can work by myself just fine."

SUCCESS AND FAILURE

There is a close connection between farmers' sense of being independent and how they understand success and failure. In much of the academic literature, questions of success and failure are treated under the rubric of inequality. Some people are more successful than others. The relevant question is why? Are those who fail somehow lacking in opportunities? Are they victims of discrimination? But success and failure are also matters of interpretation by the people themselves who are more successful or less successful.

If a person is independent, any success that comes to that person can be understood as the result of working hard and making wise decisions. But if a neighbor is hugely more successful than everyone else in the community, that poses a problem. The gracious view might be that the inordinately successful person is smarter and harder working than everyone else. A more likely interpretation would say the person somehow had an unfair advantage. Failure poses problems as well. Someone who fails perhaps did not work hard or made foolish decisions. Or there might be exceptions, especially in understanding one's own failures. Perhaps it was bad luck rather than something for which the person should be blamed.

Everybody knows people who are more successful than themselves or who have failed. Status is a matter not only of objective wealth and earnings but also of how people view one another. Farming communities are no exception. Farmers live in glass houses. Their neighbors know almost immediately if they buy a new tractor or put up a new machine shed. Gossip spreads almost as fast that Farmer X has rented another eighty and Farmer Y has purchased the quarter section across the road.

The farmers we spoke with agreed that these are the kinds of topics that get discussed whenever farmers do have time to meet for coffee or chat for a few moments at the grain elevator. They know how stories of what they are doing circulate behind their backs. Sometimes the reason for showing up at the café is to make sure stories are *not* being told behind one's back.

One farmer acknowledged that he has tried to keep his neighbors from knowing how well he is doing. He put a new machine shed behind some other buildings. It was invisible from the road. Another farmer admitted to driving an older truck than he could afford. It clearly mattered that our interviews were being conducted anonymously. Farmers talked more openly about their land and equipment than they would have with neighboring farmers.

Farmers experience another relationship that conditions them to be interested in discussions about success in a distinctive way. This is the real or perceived relationship they have with people outside the farming community. These include siblings and childhood friends who have gone on to good middle-class jobs in cities and suburbs. They also include members of the nonfarming public who make comments about farmers in newspapers and on television. Farmers perceive themselves as being the objects of criticism in these relationships and comments.

"People who are employed gainfully in industry or commerce unrelated to farming regard themselves as being the smarter members of the family," one of the most successful farmers we spoke with asserted. "It's always the simpletons who couldn't get a job in town who stayed on the farm." He added, "There is a notion that farmers aren't very smart and that they don't really have the wits to go to town and get a job."

His concern was not without warrant. For decades popular literature described farmers as country bumpkins. Farmers could hardly speak good English, in these depictions, let alone converse about literature and the arts.

Serious researchers conducted studies to see whether people who stayed living on farms were less intelligent than people who moved to cities.

This perception of being regarded by city folks as stupid played into farmers' understanding of success and how it related to being independent. They emphasized in small ways as they talked about their farming that they *were* smart and they *were* successful. They mentioned books they had read, college courses they had taken, and intricate aspects of chemistry and agronomy and economics that went into their business decisions. Being independent was important partly because it demonstrated a farmer's ability to succeed without help from anyone. Being successful involved mastering a great deal of knowledge.

At the same time many of the farmers we interviewed denied that success and failure were readily observable, even though neighbors knew and talked about one another. They mentioned two important ways of preventing success and failure from becoming as public or as closely associated with understandings of personal responsibility as might otherwise have been the case. One way involved simply keeping mum. They talked openly with us about how many acres they farmed and how much their machinery cost because we promised to not disclose their names. They said things they would not feel comfortable discussing with their neighbors. The other way was to deny that the observable marks of success and failure were the truly important ones. An outwardly successful neighbor, for example, might have debts or be unhappy.

Distinguishing true success from apparent success was evident in Mr. Rayburn's remarks about his neighbors. "I've got some neighbors who live in a nice house," he said, "but they've got a lot of debt." At least he assumed they did, although he did not know them well or see them that often. "A perception on the outside is kind of like reading a book," he added. "You can't judge it from looking on the outside."

"You can't look at something and tell," another farmer replied when asked about success and failure in his community. "You have to know them quite well to know if they are financially secure and enjoying themselves." In his view the real measure of success is whether people are happy with what they are doing.

Farmers' emphasis on individual personal responsibility is nevertheless evident in how they understand success and failure. With regard to their own successes and failures, they readily take credit for the former and usually do so for the latter as well.

While denying that it was always possible to judge success from the outside, for example, Mr. Rayburn believed that being able to acquire more land was "a pretty good judge of the character and stability" of a farmer's operation. There was a sense of justice in being able to purchase or rent more land, he felt, because neighbors ready to sell or offer their land for rent would make it available only to a farmer who had earned a good reputation in the community.

The Loeschers are an example that illustrates a similar point, but on a different scale. They have had their share of hard luck and know that other farmers in their community are more successful. They nevertheless say that how well one does is mostly a function of hard work. The land they rent is not as fertile as most in the area. They consider it a matter of pride to have cleared it and made it as productive as they have. You might be "feeling lazy," Mrs. Loescher says, but if you do not work hard, you are only hurting yourself. She considers farming different in this respect from a job in town where you could cut corners and get away with it.

Working hard in this view involves more than simply putting in long hours or doing physical labor. It involves passion. As one farmer put it, "The ones who have passion do well. They do a good job and are proud of what they do." The contrast, in his view, was people who "don't care, don't take care of their equipment, and don't care how many weeds are in the field." Those farmers, he thought, "end up falling apart."

Among farmers who had made good decisions that helped them to expand, these decisions were also in their view a mark of having thought more creatively than one's neighbors. "You walk when they are running and you run when they are walking" was how one farmer put it. Independence meant not only working harder or being smarter but also thinking outside the box.

The aspect of this emphasis on personal responsibility that is sometimes missed in the academic literature is the extent to which it involves risk-taking. Arguments against rugged individualism suggest that people who are firmly embedded in social networks and who have the backing of solid social institutions usually fare better than loners do. The assumption is that social networks reduce risk by providing a kind of safety net or communicating relevant knowledge. It is desirable to have this kind of security, according to this argument, instead of going it alone. But farmers' desire to be independent illustrates a different attitude. Part of being independent is taking risks with the full understanding that one can fail.

"I remember two or three times when I went ahead and made an investment that was going to take some time to pay off," a dairy farmer recalled. He remembered seriously questioning himself about these decisions and realizing that he was left with very little money for ordinary living expenses. Still, it was the part of farming he liked best. He enjoyed the risk involved, knowing that he could lose money, but being willing to put his knowledge and judgment on the line. "I wanted to be able to do it myself and show that I could do it," he said.

A cotton grower who had worked from being a day laborer to farming more than a thousand acres explained that he had "lost my butt" several times, but he enjoyed the challenge of taking risks. "Just one slip, man, in this day and age, good lord, oh boy!" He was not complaining. "This makes Las Vegas look sick," he said. He was pleased to have played the game and won.

"Certainly there is joy in working for yourself," a truck farmer observed. "You are as in control as much as you can be." She said the farm could go under and she would have to find some other kind of work. She would rather bring herself down, she asserted, than have somebody else bring her down. "There is no backup. You are it. Whatever you have comes from you."

A farmer like this knows it is not literally true that there is no backup. She and her husband have Social Security. They farm inherited land. Farmers in the area receive government subsidies. The point is not that being independent means being entirely on one's own. It is rather that the possibility of failure is part of the game. Taking personal responsibility means being willing to take risks.

It also means taking risks for something other than the possibility of large rewards. The risks are not like investing money in stocks that might lose money in hopes of exceptional growth. They are more like doing something with little hope of high returns at all because of the challenge involved. As a wheat-belt farmer put it, "You're probably going to harvest only six or seven of every ten crops you plant." He says you know that and take on the challenge anyway. "You can't play the game sitting in the shed."

The challenge involved, as farmers described it, is somehow related to self-realization. It can be anything from learning to play tennis to being the first in one's family to finish college. The challenge in farming varies as well. Farmers view it as doing something physically difficult. They also describe the small triumphs from making hard decisions and standing by those decisions.

"I want to build something with my hands if it takes a lifetime to build," says a hardy woman in a Hollywood movie set on the open prairie. Quoting her, a man who is barely making it on a small hardscrabble cotton-belt farm says this is his reason for farming. "That's kind of what my wife and I are doing," he explains. "We started on this little place and had some land cleared. We did some fencing and put in a pond. It's just a joy to see what you can do."

Associating individual success or failure with personal responsibility is reinforced in farmers' interpretations of inordinate success. Although they respect the neighboring farmer who has two or three times as much land as everyone else, they do not credit that person with simply having worked harder or made smarter decisions. They instead find some other explanation. This is where farmers' intimate familiarity with the family histories of everyone in their community comes in handy. They know that Farmer X had a wealthy uncle and that Farmer Y's parents struck oil.

Neil Jorgensen's neighbor had just purchased three expensive pieces of equipment, leading Mr. Jorgensen to believe that his neighbor was doing pretty well. However, Mr. Jorgensen had an explanation for this neighbor's success and for similar cases. It all had to do with "what they started from" and the "hand-me-downs" from which they had benefited. One of the other neighbors was a very successful farmer who owned a lot of ground, Mr. Jorgensen observed, but that farmer was an only child and inherited everything when the previous generation died off.

"Free money" is the term in some of the farming communities we visited that provides a shorthand way of discrediting the success farmers observe among some of their particularly successful neighbors. "The main reason I'm not farming them," a wheat farmer who farms about five quarter sections less than he thought about trying to farm at one time explains, "is because it takes money. And I have never gotten what I call free money, which would be inheritance. I've had to earn everything and I've always had debt."

One additional connection between success or failure and farmers' sense of independence comes into focus in relation to the hardship stories that are so common in farmers' narrative repertoire. In the same way that traditional rags-to-riches stories highlighted the special virtues of those who succeeded, rural hardship narratives sweeten the meaning of having made the right choices.

Stories of overcoming hardship can be told in any line of work. Among doctors the stories may emphasize the grueling experience of medical school. For teachers, law enforcement officers, and small business owners, years of low earnings provide grist for hardship stories. Farmers have the additional experience of wild fluctuations in income due to bad weather and shifting market conditions.

One of the dairy farmers we interviewed recounted the difficulties he and his wife experienced during their early years of farming. The worst part was that the barn burned down. Besides the barn, they lost the majority of their herd. The setback almost convinced them to quit. They kept the dairy going by working at full-time off-farm jobs. Then interest rates went through the roof. He is proud to have survived these difficulties. Having prevailed vindicates the decisions he alone took responsibility for.

Noting the connection between overcoming hardship and valuing independence does nothing to minimize the fact that many farmers have triumphed over hardship. The ones who have survived in farming know of relatives and neighbors who did not. They do not feel left behind or unsuccessful because they are still on the farm. Instead they consider themselves survivors, the one among their siblings, the few among their classmates who overcame difficulties, worked hard, made the right choices, and had the passion necessary to succeed.

Feeling beleaguered by bad weather, adverse market conditions, overbearing agribusiness companies, and aggressive neighbors means, in farmers' narratives, that only the ones who truly work hard and make wise decisions succeed. That kind of independence involves acknowledging the forces shaping one's destiny. It means that whatever freedom a person does have must be used with great care.

PERSONAL RESPONSIBILITY

The irony about farmers' emphasis on personal responsibility and independence is that many of them did not consider independence a trait to be encouraged early or strongly in young people. They thought independence was something that happened in cities and created problems. Young people in cities were into doing their own thing. They had too much freedom. Too little supervision.

This concern is illuminating. It helps interpret how farmers' understand independence. The reason they consider it desirable to be their own boss is that they assume they know what it means to be their own boss. The rules are clear. They are well institutionalized in the community and in the practices required to succeed in farming. The principles involved have been internalized. A boss is in the first instance self-*governing*, not a person who feels free to do anything and everything.[7]

The kind of independence they want young people to learn is principled. It requires both the technical skills that growing up on a farm may have taught, such as driving a tractor or baling hay, and the judgment that comes from having practiced those skills. Judgment is harder to assess, but emerges in farmers' complaints about offspring and neighbors who do not exhibit it. Judgment involves taking responsibility for one's decisions—doing the best one can and living with the consequences.

That was the connection farmers we spoke with saw between being independent and being a person of strong and decent character. Taking responsibility over an extended period for one's decisions required a level of personal fortitude, a degree of commitment to carrying through on one's decisions that they felt was especially evident in farming.

The notion of farming as a practice is, again, evident in this understanding of independence. A person who becomes especially good at something from having practiced it, whether the task is farming or learning to play the piano, assimilates a set of rules. Being able to improvise may be part of the game, but improvisation is possible only because of having first learned the rules of the game. It takes discipline to master the skills involved.

The trouble with being independent, indeed, is that everything comes down to having internalized those self-guiding and self-governing principles. Any successes achieved are one's own, but so are the failures. Personal responsibility can thus be a heavy burden. "Sometimes being your own boss is not good," a farmer told us at the end of a long day on the tractor. "Nobody's telling you *not* to go out there and work."

This is the aspect of personal responsibility that critics of American individualism have emphasized. Sure, it is important to be free and to take responsibility for one's decisions, the argument goes, but there have to be limits. Individualism has to be kept in check by social relationships, sharing, and taking into account the needs of others. Otherwise, it leaves people too much on their own, uncertain of how to govern their own lives, and insufficiently attentive to the value of working together.

It was clear among the people we interviewed, though, that emphasis on personal independence in farming communities is subject to social norms. If people grow up wanting to farm because they can be their own boss, they also learn about responsibilities to other people. As we have seen, those responsibilities are reinforced by strong loyalties to family and by understandings of neighborliness and faith.

The social role of taking responsibility for oneself is partly achieved through behaving in ways that avoid burdening the community. A farmer whose cows get out too often becomes known as a poor farmer. Keeping one's business dealings to oneself is better than being known as a braggart or the village gossip.

The norms governing independence identify persons considered *too independent* and subject them to local criticism, or worse. In local parlance these were the farmers their neighbors poked fun at or held in mild disdain. Idioms identified them as the kind of person who thought they invented the tractor but probably did not know how to fix one. They were too bullheaded to take advice. They were like the proverbial hog on ice, trying to be independent, but awkward, sliding around, insecure.

No matter how self-sufficient a farmer was expected to be, it was expected that some information and advice would be shared. It was anathema, farmers said, to be blatantly secretive. If a neighbor asked what kind of seed you had planted or where you thought the best deal on a used implement could be found, you were not expected to say, "That's none of your business." You could lose friends if you did. Or you would be passed over when land became available for rent.

The farmers we spoke with were uncertain whether being truly independent, in the best sense of the word, was an ideal that could still be realized in farming or whether that possibility was declining. They knew that farmers were always at the mercy of weather patterns and market fluctuations. Compared with working in a factory or holding an otherwise attractive job at a large company, farming permitted a person greater autonomy. It provided opportunities to make one's own decisions and to do different things from day to day. The downside was that rented land put farmers under the control of landlords and expensive equipment turned farmers into servants of the banks.

Feeling that they were no longer as independent as they once were was especially rooted in the fact that farming was becoming more highly specialized and more technologically sophisticated. A large tractor that enabled a farmer to do more work without hiring someone to help nevertheless was a tractor that could not be repaired easily. "Twenty-five years ago if we had a tractor that needed to have the engine or transmission overhauled," Mr. Rayburn observed, "we'd throw it in our own shop and do it. Nowadays we don't." He still enjoyed being his own boss, but recognized that this was a way in which he was less independent than he had been in the past.

One surprising argument, though, was that farmers are actually becoming *more* independent. The logic was that farming on a larger-scale with better equipment gave farmers an edge in withstanding market fluctuations. They could diversify, protect themselves better from an occasional hailstorm or crop disease, spend more time managing and innovating, and still do most of the work themselves.

Clay Jorgensen was one of the farmers who held this view. "I think the younger generation is more independent than farmers our age," he said. Why? "They seem to understand things more." He meant the benefit of having more education and being able to farm on a larger scale because of the information at their disposal about technology, chemicals, and markets. He wasn't sure, but he also thought maybe farmers were less dependent than they used to be on their families and neighbors. He could see that Neil and Arlene were becoming more independent. Maybe that was a trend.

Some of the other farmers we spoke with described a similar trend. They thought younger farmers in their community were behaving differently from younger farmers a generation ago. But there was a disturbing aspect to this new form of independence, at least as older farmers described it. Being independent seemed to imply a greater willingness to undercut one's neighbors. It meant less willingness to share ideas and more emphasis on self-interest.

Farmers held differing opinions about how significant these changes might be. They knew from their own experience that working hard and taking personal responsibility for their decisions was preferable to following the crowd. And yet it bothered them to think that a spirit of competitiveness was replacing an ethos of community.

In the final analysis, though, farmers' understanding of independence is less about their relationship with other farmers than with themselves. Independence involves strength of character. And character is built by taking on the challenges of making hard decisions, taking personal responsibility for those decisions, and seeing through whatever it takes to realize the consequences.

Character develops in the same way that it has to for anyone in any line of endeavor through practice. In farming its development is facilitated by the circumstances of being one's own boss and thus of having to live with the consequences of one's decisions. A cattle farmer in his fifties provided an apt summary of this idea when asked what values were especially important in farming. "I want to say courage," he began, "but I don't know if it's as much courage as fortitude."

He added, "Things don't always go just the way you thought they were going. You wonder if you made the right decisions. You need to start on something and stick to it, but not get rigid about it. You have to be versatile and reassess where you think you are going."

THE LAND

5

When you work the land for so many years, you identify with it. It is something we take care of. We're stewards of it.

—Corn-belt farmer, male, age 66

I'm taking care of it for the next generation, whoever that may be. The day is coming when I'll be in a nursing home. I want them to say, "you know, old man, you did it right."

—Wheat-belt farmer, male, age 57

"Here's my field," Bill Sellers says, pointing across the open expanse as he pulls the pickup to the edge of the road. "I just want to take a quick peek, see how ripe it is." He walks a few paces into the field. The wheat is knee high but still green. "This is one of my best pieces of ground," he says. He grabs one of the stalks, opens the head and squeezes one of the kernels between his thumbnail and forefinger. "You want it to be hard dough," he says. "That head has just a little bit of juice in it, which may be alright. Once it gets into hard dough, then you can spray it."

Mr. Sellers has been inspecting the wheat in this field for nearly half a century. He grew up just down the road. As a boy he watched his father cultivating the field and harvesting the wheat. In those days you watched and waited. When the stalks turned color and the kernels hardened, you harvested.

Things have changed. Mr. Sellers spends far less time in the field than he used to. But he has to know a lot more to get the best crop. The wheat is modified so weeds can be sprayed at a particular moment without damaging the wheat. "If you spray it too soon," he explains, "it will shrivel up the kernel and hurt the quality of the wheat." He decides to wait a few more days before spraying.

"Look at this," he says, examining another stalk. "You can see some streaking there. That's a little bit of disease." He points to a black spot near the top of the head. "That's a scab disease. This year we put fungicide on. That helps prevent it." Without the fungicide, more of the heads would have been empty. He figures the field will still make forty bushels an acre. In the old days, fifteen would have been good. Now forty is below average.

What does the land mean to a farmer like Mr. Sellers? Does it still hold the kind of attraction that drew the first settlers—the ones who broke the sod and staked out a place for their families? Does it mean what it did for Mr. Sellers's father after spending long days each summer plowing the field? Or has it become a place in which the technical expertise involved in spraying weeds and applying fungicide is the main preoccupation?

Land, labor, and capital. In economic theory these were the classic factors of production. Their utility was to produce goods. The story of economic modernization was a shift from the first to the third. Peasants labored on the land, earning surplus value in small measure for its owners. The modern entrepreneur accumulated capital and built factories and made labor so much more productive that the land hardly mattered.

In the modernization story the few farmers who remained did so by following the entrepreneurs' example, making the land more productive by applying capital. Human labor gave way to horse-drawn machines that in turn gave way to engine-driven tractors. Those became larger and more powerful until chemicals dwarfed even their share in the expansion of productivity.

The missing element in the modernization story was that the land was always more than one of the factors of production. A financier might have felt a deep emotional attachment to a bank account and some factory owners were known to feel that way about a steel or cement plant. But those kinds of attachment were probably rare or at least unexpected. It was different for farmers. They were known to be sentimental when a favorite horse or cow died. They also developed an attachment to the land. They lived on it. They cherished it because of the stories and the memories they associated with it. It was family land. It was their home.[1]

The resulting attachment was best described as ambivalence. Land was appropriately an entity to which sentiment could be attached. A farmer could take pride in farming land that had been in the family for generations. A person who spent days on end tilling the fields could develop a kind of intimate respect for the terrain. At the same time the land was still a factor of production.

Mr. Bower, the wheat farmer we have heard from in previous chapters, captured contemporary farmers' ambivalence toward the land they farm as clearly as anyone with whom we spoke. "I know the ruts of the old wagon trail that goes through the pasture over there." He has a deep appreciation for the people who farmed the land over the years. "But I also have this rational detachment. It's dirt and this is a business."

Describing farmers' relationship to the land as a trade-off between romantic or nostalgic conceptions and hard-nosed business considerations, though, is an oversimplification. For many farmers the fact that land has been in the family for several generations is only one of several important ways in which the land is meaningful. As individual farmers farm more land, less of that land is likely to have been in their family for a long time. That difference has implications for how they think about it. No matter how long they may have farmed a particular piece of ground, the most immediate relationship to the land is working it. An outsider might describe a field abstractly as a patch of wheat or corn, but a farmer knows the field more intimately.

Those relationships too are changing, though, as technology and machinery change. The somatic connection with the land that comes from working the soil by hand and breathing the dust is diminishing. Escalating investments in mechanization are challenging the family connections with the land that earlier generations considered beneficial for the development of strong character. The mental relationship is changing as soil science offers nuanced understandings at the molecular level. As we shall see in this chapter, farmers' relationships to the land are also increasingly shaped by questions about sustainability and by considerations of who has the final say in addressing those questions.[2]

Listening closely as farmers talk about the land they farm and what it means to them reveals the complexity of these relationships. Farmers are perfectly aware of the changes that are requiring them to farm more land, use larger machinery, and rely more on advanced technology. They understand the costs as well as the benefits. At the same time their thinking cannot be interpreted simply as a full-throated embrace of large-scale industrialized farming or as a sorrow-filled tale of regret for better days. Faced with the daily challenges of the life they have chosen and are still committed to, they stitch together ways of honoring the land and making sense of their efforts to preserve it.

FAMILY LAND

Mr. Sellers settles behind the steering wheel of the pickup again. "I just want to drive around and take a look at the east end," he says. As he drives, he talks about the land. The field is a mile on each side—640 acres. As far as he knows, it has never been subdivided, which is rare even in this part of the country where fields are large and farmsteads are few. "There aren't many square sections left that have all been farmed in one chunk for probably a hundred years," he says with a hint of pride in his voice. "My dad said, 'This is never going to get broken up. You remember that.'"

"I've lived in this spot all my life," Mr. Sellers continues. "Our old house used to sit right there," he says, pointing to a low ridge that just appears to be an empty field. He was in grade school when his dad built the house, and after his parents moved to town he and his wife lived here until a few years ago when they built a new one nearby and had the old one moved away.

The land itself was in the family at least by the end of the nineteenth century. Mr. Sellers says he has forgotten for sure if his grandfather or a great uncle or maybe even someone of the previous generation owned it. What he knows for sure is that his father and mother started renting the land in the 1950s—he gives the exact date—from the relative who owned it, moved into the old farmhouse that had been there since the 1930s, and then purchased the land when the owner died.

"Here we are," he says, pulling the truck to the edge of the road again. "You see here on this end it's lower." The difference is almost imperceptible. "See this whole east end. That got drowned out. The water all runs to the east end and that's got to get out into this ditch. Sometimes the ditches get full, so it got hurt. It drowned out quite a bit of stuff here in the east end. That's just the way it is." Next year he plans to plant corn and soybeans here instead of wheat. "They take the water way better," he says.

The kind of intimate relationship that Mr. Sellers has with this field is typical among the farmers we talked with. They know the terrain in as much detail as if they still walked it with a team of horses like their ancestors did. If the land has been in the family for several generations, they know its history. The land is part of their family story. They do not have to visit the county courthouse to recall who owned it. The previous owners are people they knew and loved.[3]

Having been in the family means that land is the occasion for memories and stories. See that shelterbelt at the edge of the field. Grandpa planted it in the 1930s. Aunt Tilly lived here where you see this clump of trees after Uncle Fred died. This piece of ground belonged to our neighbor when our parents lived down the road. He was a bachelor and came over to visit on Sunday afternoons. He disappeared one day. Nobody ever saw him again.

A truck farmer in his late fifties who farms about a hundred acres that have been in the family for three generations had just been thinking about the connection between land and family stories when we talked with him. He had recently given a talk about it to a group in town. "I can take you to any field on this farm," he told them, "and I can give you a story. I can tell you about my grandfather. I can tell you about my father, something that happened with my mom, and something that happened with my kids, where my wife and I went parking when we were dating. Every field has a story."

Few of the stories, he said, are the kind that would make it into a novel. It is rather their ordinariness that makes them meaningful. One of the fields back through the woods always reminds him of his grandfather. It's called the swamp. He was twelve when his grandfather asked him to walk with him to the swamp to watch his father mowing hay. "It was nothing in particular," he recalls, "just being with him and my father."

Another field brings an image of his mother to mind as clearly as if it were yesterday. She was driving the tractor pulling the hay baler while his dad took a wagon of hay bales to the barn. When his father returned, he saw that the baler had run out of twine. It was kicking out untied hay squares rather than bales. "Didn't you notice the baler wasn't working?" his dad called out. His mother retorted, "Damn it, if you don't like the way I'm doing it, do it yourself." It was just the simple things like that, the man said. They made the land special.

One of Mr. Sellers's neighbors is old enough to retire. The man's grandfather homesteaded the land in the late 1800s. At first the man denies he has any special attachment to the land. Right now he and the land are not getting along. The land is soggy. Some of the corn is standing in water. But his wife chides him. She thinks they should move to town and let their son take over the farm. She says her husband is holding out. "I think he does feel attached to it," she says, "because he does not care about moving to town."

For older farmers like this couple feeling attached to the land is only partly the kind of inertia that anyone feels who has lived in a particular house or community all their lives. They remember—or have heard—the stories of hardship that went into keeping the land. Their own efforts to hold on to the land are a way of honoring the sacrifices made by their parents and grandparents.

After his wife scolds him the man tells with some pride how his grandparents struggled to keep the land during the Dirty Thirties. Nothing would grow except thistles. His grandparents tried to cut the thistles and feed them to the cows because there was no hay, but the cows balked. In the 1940s things were still difficult. Prices were higher, but rationing during the war made it hard to buy enough gasoline to run the tractor. The man's father worked for the railroad to supplement the meager earnings from the farm.

The farmer I mentioned in chapter 2 who exports purebred livestock to Brazil and Argentina is similarly attached to some of his land. Although he has purchased thirty farms and accumulated more than ten thousand acres over the past four decades, he especially values the one small farm that has been in the family for four generations.

He knows the stories of his ancestors farming with horses and struggling to keep the farm during the Depression. The land is special because it is intertwined with memories of a great grandfather who came as an immigrant, a grandmother who died giving birth, and a father and uncles who worked hard on the farm all their lives. "I want to be buried on this farm," he asserts. "It has special meaning."

One of his neighbors who farmed on a much smaller scale felt the same way. The land was dear to him. It was especially dear because he had been working it for four decades and because he had lost some of the neighboring land he formerly rented. "I told my wife that when I die," he said, "I want to be cremated and I want [the ashes] to go on a plane and just be thrown out over this quarter section here where I live."

Remarks like this connect with a long history in American culture of feeling a special attachment to a particular place, of considering one's place in life as a physical location, and of wanting to claim, own, and cultivate that place. They run counter to the equally prevailing theme in American culture that encourages people to leave home.

Being attached to a place can also be interpreted as a source of imprisonment. That was true for some of the farmwomen we spoke with. They felt tied down, enslaved, and indeed left behind. Living on a farm felt isolating, limiting, a hindrance to other career opportunities they might have explored. The more common view, though, was one of having made peace with one's location and of feeling that things were good in this particular spot, perhaps better than they could have been elsewhere. The farm was a secure, familiar place. It was home.[4]

The land often conjured up feelings of nostalgia as farmers recalled parents, grandparents, neighbors, and methods of farming that no longer existed. It was relatively free of homesickness, though. Although a few of the farmers we spoke with said they felt lost when they traveled, most enjoyed getting away from time to time. Being relatively freer to come and go than they imagined farmers had been in previous generations kept the land from seeming like a prison.[5]

Like family members, the land is never perfect. It seldom lives up to one's highest expectations. You make the best of it. A fourth-generation farmer recalled her father's frustration when the land failed to produce: "He said you could make more money drilling assholes in wooden hobby horses!" Mr. Sellers wishes the east end was not so low that the water backs up and drowns the wheat, but he works around that problem.

A couple we spoke with who raise cattle on the grassland they farm had just the opposite problem that Mr. Sellers did. They were experiencing a severe drought. Annual rainfall in their area averaged eighteen to nineteen inches, but in the past year was less than six. The man said they were struggling to water their livestock. The wife said they had battled wildfires. Some of their fence had burned as well as the grass. But the man added, "I literally love the land." His wife agreed.

Apart from problems like these, the trouble with family land is that inheriting it is nice but not quite the thing that gives a sense of personal accomplishment. Land that you did not inherit—land that you purchased—is more meaningful in that respect. Even land that you have been able to rent can be a source of pride. You were able to rent it because the landowner trusted you. You got it instead of some of the neighboring sharks who wanted it because you had earned a reputation as one of the best farmers around.

Mr. Sellers illustrates this kind of attachment when he talks about the land he farms in addition to the section he inherited. He currently farms five or six times as much land as his father did. "Land is tough to get," he says. "You get a crack at some land to rent, man, you better give it your best shot because it probably will happen one time in your life that a neighbor's land will come up for rent." To get that land, he says a person has to be competitive, even aggressive, and willing to take some risks, but mostly it matters to be known in the community as an excellent farmer.

Farmers who "just slide into the farming operation" by inheriting all their land, Mr. Sellers thinks, have it too easy and sometimes are not the best farmers. "They've got it made," he says. In this respect he takes pride more in the fact that he has been able to rent and purchase land than in the part he inherited. Most of the additional land came from neighbors who saw he was a good farmer and decided he was the best person to take over their land. "My dad didn't really have the means to help me a whole lot," he says, "so I started all on my own. I guess, to me, that's probably the thing I like to think about, what I've done."

Mr. Lancaster, the wheat-belt farmer we met in chapter 1, shares a similar view. Although he benefited from land his grandfather acquired and passed to his father, he feels especially rewarded by having acquired additional land. Even if the land is rented, the fact that it is scarce and has been offered to him instead of someone else makes it special. Neighbors have come to him and said, "Hey, we're done. We've watched you for the last number of years. We want you to farm our ground." He says "there's such huge satisfaction" when that happens.

The missing piece in such narratives is that family connections may still be important, and indeed often are when more of the story comes out. The land does not have to be inherited for that to be true. Many of the farmers we spoke with agreed that hard work and having a good reputation in the community were not enough. Unless a farmer was related in some way to an extended family with large land holdings, the chances of being able to rent additional land were low.

"Around here, if you're related to him," another wheat-belt farmer says in describing one of his wealthier neighbors, "you have an in. But if you're not related to him, then your chances of finding farm ground are just null and void."

The way he says this makes it sound like he is complaining. He is, to some extent. But he acknowledges that family connections have played a role in his own ability to acquire land. His wife's relatives own some of the land he rents. He was able to rent some other land because the owners were related to his mother. "The biggest thing as far as getting farm ground," he says, "is being related."

Even if it is inherited or acquired through connections with relatives, though, family land is still special in a way that land is not if it is obtained in the wrong way. The right way is to inherit it or to earn the right to rent or purchase it by working hard. Nobody can fault the neighboring farmer who purchases or rents additional land as a result of hard work and good management. Land speculators are different. So are outside investors. And so are people in the community who acquire land through shoddy business practices.

Neil Jorgensen had been talking for a while when he gingerly broached the subject. There was a lot of bootlegging going on in a farming community to the south of where he lived, he said. It was common knowledge that this had been going on for a long time and that some of the farmers there were making good money selling bootleg whiskey. Some of it was just being made for personal consumption or to give as Christmas gifts. But some of it was selling for as much as $500 a gallon. Thousands of gallons were being sold at that price. "Those people own land like you wouldn't believe!" Mr. Jorgensen exclaimed.

Whether it was bootleg cash or money coming in from speculators, the difficulty was that outside money was sullying the land. Prices were being driven up artificially. That might be good if a farmer was planning to sell out and move away anyway. But for those who planned to stay in farming, it seemed morally wrong. It violated the basic rules of the game.

WORKING THE LAND

Besides owning or renting the land, the relationship that gives farmers the special and most immediate relationship with the land is working it. The meaning that results is quite different from the one a person derives from walking leisurely through a public park and thinking what a pretty place it is. The meaning a farmer attains is closer to the satisfaction a homeowner gains from painting the house and tending the yard or that a musician acquires from performing an original composition.

Making it look good. Making it look like your garden. Making everything as pretty as you can. These are farmers' ways of describing their relationship with the land. The relationship is in no small measure aesthetic. Farmers emphasize the beauty of fields turning green from recently planted wheat or corn. They value the look and smell of a newly plowed field and the vista of ripening grain undulating in the wind.

The aesthetic appeal includes the quiet serenity of living in the country. For people who appreciate solitude, relating to the land is calming in a way that being around people is not. "It is comforting," a woman who grew up in a small town in the cotton belt and has now lived on a farm for three decades explains. "When we moved out here, it seemed isolated and people asked if I was afraid. No, you just walk outside and it's black dark, and there's something about the serenity and the peacefulness of the land."

The calming presence of the land is so constant that she seldom has to think about it at all. That was true of one of the wheat farmers we spoke with as well. He is more intentional about making time to appreciate the land. He views it as a way of relating to God. Planting wheat gives him a sense of doing what God wants him to be doing. He sets aside time to think about that relationship. Some evenings he drives his four-wheeler to one of the fields. "I just listen to the crops grow," he says, "listen to nature, take time to meditate, soak it all in and just enjoy the beauty that has been given to us."

These aesthetic relationships defy the notion that farmers are just money grubbers who will do anything to make the land more productive. The beauty of the land and farmers' emotional attachment to that beauty are important too. They are enriched by virtue of the labor involved. The seed that is now sprouting through the soil had to be planted at the right time. Part of the beauty is that the rows are straight. The correct seed and fertilizer have been used. An understanding of the soil's strengths and limitations is necessary.

"Rather than go fishing," a farmer who raises cattle muses, "I'd rather go out and look at the pasture and see the cows out in the pasture. Count the cows. Put up hay. I love to put up hay." He mentions the smell and the challenge of drying it and baling it just right.

"One of my greatest joys in farming," a third-generation corn-belt farmer notes, "is watching the seed, the new plants come out of the ground. It's new life. And calving time is special for me. I just love seeing the new calves. No other job can take its place. It's new life, new birth every year. It's just awesome."

A cotton grower who annually plants about three thousand acres said the part of farming he likes best is seeing the plants about three inches high each May or in early June. "Just having that green carpet across the field where we haven't seen anything green since September—that's a beautiful sight." Not only that. He then gets busy babysitting the plants. "You can't let the sand blow. The sand particles will blow across the ground and cut into the little tender shoots. So you hurry and run the sand fighters. They have spikes that turn up a little fresh dirt so you don't have that slick sandy-looking surface."

The land is the resource from which pride of accomplishment derives. Although the land is in many ways unchanging, it is the silent partner whose cooperation must be secured for everything that does change. When farmers describe the meaning land holds for them, it is seldom as an inert abstraction. What they enjoy is seeing things grow, working with animals that depend on the land, and working to make the land better.

Here at the east end of the field where the water collects, Mr. Sellers is thinking about what should be done. "We've got to deal with Mother Nature," he muses, "but we have no control over her." He finds it hard to strike the right balance between what he cannot control and what he can control. When it rains he knows he might as well take the day off and go to the lake or repair machinery in his shop. But having the water drown the crop here in the low end of the field seems different.

"We're in a wet cycle," he says. "We live in the bottom of [what used to be] a lakebed. This was the bottom of the lakebed years ago. So we're probably going to be a little wet." He thinks the time has come to invest in tiling his land like many of the farmers in his area have already done. "Tiling works great," he says. "Ground that's tiled is like a sponge. When you squeeze the sponge out, that's what the tile does. It squeezes it out slowly, and then when you let go of the sponge it's sitting there, but then if it rains again, it'll take some more."

Tiling is clearly a major business decision for Mr. Sellers. It will cost between $600 and $700 an acre to install the underground tubes thirty or forty feet apart plus some culverts and pumps. As a business proposition, he figures it will increase annual productivity and add to the value of the land. He is also personally excited about the project, though. It makes him feel good to be doing something that improves the land.

Paul Freeman's farm is about as different from Mr. Sellers's farm as one can possibly imagine. Half a continent away, the fields here are small. If Mr. Sellers's fields were placed side by side and were half a mile deep along one side of the road, a person could drive for almost fifteen miles before coming to land someone else farmed. By that same measure it would take less than half a mile to pass Mr. Freeman's land. Instead of wheat and corn, he specializes in root vegetables, such as carrots, and other fresh market vegetables, such as sweet corn, tomatoes, peppers, broccoli, cabbage, and cauliflower.

Mr. Freeman says farming is "in my blood." He was "one of those lucky kids who wound up doing what he wanted to do growing up," he explains. But the land he farms is not land that has been in his family for generations. His grandparents ran a dairy farm an hour away. They sold it when he was a boy. He nevertheless thinks that "planted the seed in my head that a career in agriculture was what I wanted to pursue." With no family land and with no skills from farming while growing up, he became a farm laborer when he was eighteen. After twenty years he saved enough money to purchase his own farm.[6]

Mr. Freeman cherishes the land because it took him so long to be able to buy it. But that is not the only reason. He values its history in the same way that farmers with ancestral land do, except in his case the history refers to the valley itself and to the Dutch and German immigrants who settled it three centuries ago. During the American Revolution, he says, this area was known as "the bread basket" because it supplied the food for Washington's troops.

The land here is especially fertile. Mr. Freeman says the topsoil in most areas where farming is done ranges from six to ten inches, but here the topsoil is fifteen feet deep. During the glacial age, it got that deep by being pushed into the valley from the surrounding hills. That suits it especially well for root crops. "It's one of the best soils in the world," he observes. The area is beautiful as well. "Every day I get up and I walk outside the house," he says, "and I look across the valley here and I look up at the mountain that borders it, and I just have to say, 'How many people get the chance to do this?'"

A person who lived in the valley and commuted to work in a city an hour and a half away might say the same thing. They would probably appreciate the area's natural beauty as well. The difference in Mr. Freeman's relationship to the land comes from working it. Spending time with it day after day and year after year, learning its contours by heart and understanding what grows best from season to season is how he has developed an intimate relationship with the land. That was the relationship he developed with the land where he worked as a farm laborer for twenty years. "I felt as though it took me twenty years to understand that soil and get to know it."

Now that he has worked the land on his own farm for two decades, he feels the same way about it. "I feel well along the way on this farm and this soil," he says, "but it is a learning process." For example, the soil behaves differently during wet seasons and dry seasons. "What's it like when it's cold, how does it perform, how does it deal with this crop or that crop, what can you get away with here and there?" He says the relationship a farmer has with the land is "definitely an intimate relationship. . . . By no means is it just a piece of land."

The physical labor involved in hoeing the weeds and harvesting vegetables is an important part of the relationship a truck farmer like Mr. Freeman develops with the land. He knows what the soil feels like and how the color reveals the varying mineral content closer to the river and toward the mountain. The warm soil smells fresh in the springtime when he cultivates it, and he can sink ankle deep in it in the fall after a heavy rain.

The somatic relationship with the soil includes moods prompted by the changing seasons. The moods are different on warm sunny days than on dark rainy days, accentuated by the fact that the tempo of work changes as well. "When things start melting and it gets to where you think spring is coming," a farmer in the corn belt explains, "it's like there's a switch. All of a sudden you go from kind of relaxing to 'boy, here we go again.' It's just an excitement that I can't explain."

The relationship with the land is also mental. Mr. Freeman offers an interesting illustration. He knows the land is better at producing some vegetables than others, but he also knows that it makes no sense to grow something if there is no demand for it. So each year before the growing season begins he sits down with his three grown children who farm with him and he asks, "What six things are we going to do differently this year?"

The physical part of growing a potato stays the same from year to year, Mr. Freeman says, but the mental part has to change. He considers it insane to keep doing things the same way. Even though the land is the same, his relationship with it changes. "So each year we look for six things that are substantially different in our business. Can we grow a new crop? Can we put an addition on the greenhouse? Can we get a new piece of machinery that will make us more efficient?" Those are the questions that keep his thinking about the land fresh from season to season.

Farmers' mental relationship to the land is especially evident in the thinking that goes into confronting the challenges that different seasons and varying weather cycles present. The land in these respects is like an animal. It has to be understood to be trained. The challenge is doing that in ways that honor and preserve the land.

"See that field over there," Lloyd Johnson says. He is a third-generation farmer in his late fifties who plants more than a thousand acres of corn and soybeans annually. He points across the road from the front porch where he is sitting this evening. The field is a quarter section, one of several he owns. "You see how yellow it is," he says. "It's got a bad case of ragweed." He planted corn there early this year expecting a warm season and an early harvest. Instead it was cold and wet most of the summer. The corn didn't start. The weeds did.

This is the kind of challenge that puts Mr. Johnson's mind in high gear. He doesn't want to apply any more weed killer than he already has. His plan is to show the weeds who's boss next year. "I'm going to plant that to rye," he says. "That will help choke out the weeds this fall. The rye will be growing next spring. We'll harvest it in late May or early June and then plant corn." He figures that will cut down on the weeds and minimize the need for herbicides.

"You want to make your soils better," he says, observing that any farmer would be foolish not to improve the soils. "I want to put rye there because it will do that. I want to use minimum tillage. It maintains the cell structure." He adds, "Your soils are your assets. True and simple. The better you treat them, the better response you're going to get."

In most such decisions the tension between economic and aesthetic considerations is minimal. A weed-free field is not only more productive, it also looks good. The joy of watching new plants come to life in no way diminishes the anticipation of an eventual bumper crop yield. That does not mean the aesthetics are always pleasing. Any farmer who has spread manure knows that.

Where the aesthetic aspects came most clearly into conflict for farmers we spoke with was in issues related to recreational uses of the land. Many of them enjoyed hunting. The land was good for hunting pheasants and quail. Their friends from town liked to come out and bring their guns. A few had farm ponds that were good for fishing. Some boarded horses. Several of the truck farmers were involved in agritourism. They hosted berry picking weekends and fall harvest festivals. One of the wheat-belt farmers we spoke with had done the same. She and her husband grew pumpkins for sale, exhibited old-time farm machinery, created venues for family reunions and weddings, and provided educational tours for schoolchildren. Last year they hosted ten thousand visitors.

The conflicts occurred when these activities brought people to their communities who farmers felt did not truly appreciate the land. They worried that good pastureland was being purchased at high prices by urban investors who planned on doing nothing better than bring their rich friends to the area to hunt. They found gates left open and crops trampled. New neighbors who lived on small parcels of land and did not farm were a mixed blessing. They liked being in the country and yet had little in common with farmers who had earned a living there for generations. They did nothing but complain when manure spreading season came.

DISTANCE FROM THE LAND

Many of the farmers we spoke with acknowledged that they spend less time on activities that put them in close contact with the land than was true a generation ago or even when they began farming. A piece of ground they farmed twenty miles from home was one they saw less often than the land behind their house or across the road. A field that took three days to plow in the 1960s could now be cultivated in half a day. The dust they used to breathe while plowing now remained outside their air-conditioned tractor cab.

"If you were lucky," a cotton-belt farmer who started in the 1950s recalled, "the tractors had a little umbrella that hardly blocked the sun. But as far as dirt and wind and rain were concerned, you had to deal with it." In those days he stripped the cotton by hand and pitched it into the wagon with a pitchfork. The land was literally at hand. It was impossible to escape it. On windy days the dirt blew into the house. His wife put wet towels over the windows in hopes of keeping it out.

A farmer in his early fifties who annually plants about 3,500 acres of soybeans and 1,500 acres of corn says he misses spending time in the fields. He puts in a few hours most days when tractor work is being done. Much of that time is spent doing business by cell phone. Most of his workday is consumed with paperwork, repairing machinery in the farm shop, or running errands to town. "I don't get to do the things I really enjoy about farming like I used to," he says.

"When I was eighteen years old," a wheat-belt farmer says, "all you did in the summer was get up, get on a tractor, stay there all day long, and go back home at the end of the day." He was literally in the field close to the land every day for days on end. He now practices no-till farming on the 3,000 acres he farms. "We have ground that we haven't actually tilled going on twenty years," he says.

A dairy farmer who owns 350 acres and rents another 1,500 grew up lifting hay bales and clearing stones from the field by hand. His relationship with the land in those days was physical. It included touch and smell. Aching muscles. He still enjoys watching "nice green crops grow" and harvesting good crops. He says he would never sell any of his land for housing developments. But his day-to-day activities involve hardly any direct contact with the land. He spends half the day in his office paying bills or checking water levels and temperatures in the barn. The smells of manure and new-mown hay are still part of his experience, but he does not perform the labor himself.[7]

Truck farming also provides an interesting case in point. Although most of the truck farmers we spoke with still walked the land regularly and took pride in what it grew, they also noted a significant change from what they remembered growing up. Mr. Granger was one such case. He recalled walking the fields picking up rocks and hoeing weeds. Now he spends less time in the fields. The land he farms is in three counties. Some of it is thirty miles from where he lives. Some days his machinery spends more time on the road than in the fields.

Another truck farmer we spoke with cautioned us from presenting a picture of farmers' relationship with the land that was "too pretty and romantic." Although she and her husband were farming land her grandfather acquired in the 1930s, she viewed the land chiefly in terms of its contribution to the family business. She spent more of her time in the farm office handling paperwork than doing work in the fields. Much of the business involved contracts with smaller farmers in the area who actually grew the corn and root vegetables.

Among farmers whose relationship to the land was becoming more distant in these ways, an important result was that their understanding of the land was less visceral and more conceptual. The land was still a matter of bodily experience to the extent that it felt a certain way when walking across it and looked and smelled in familiar ways after a warm spring rain or during harvest. The embodied relationship to the land, though, was less prominent in farmers' descriptions than accounts of its size, history, ownership, and use.

An illustration was evident in Neil Jorgensen's comments about the acreage he farms. Although some of the land has been in the family for five generations, most of the land he farms has not, and he spends less time on it thanks to using larger machinery. He has also become a fan of no-till farming, which he believes conserves the soil. "We're getting back to originally when the Indians first came," he says. This is a historical reference, a concept that suggests the depth with which he values and respects the land and yet is an abstraction. It says less about his immediate day-to-day physical engagement with the soil and more about how he has come to understand it intellectually.

The contrasting view is evident in the ways in which Mr. Jorgensen's parents talk about the land. Clay Jorgensen emphasizes conservation, too, but emphasizes "enjoying the land" and being interested in the trees, water, and air as well as the land itself. Mary Jorgensen says she and Clay "love" the trees and "like to take good care" of the land. Their reference point is the farmstead and the pasture and cropland immediately surrounding it. Neil Jorgensen values those aspects as well, but they do not surface as readily in his thoughts.

The subtle shifts in farmers' understandings can be described partly in terms of the difference between particularistic and universalistic relationships. A "particularistic relationship" is the kind a person has with a parent, child, or spouse. A "universalistic relationship" refers to a person's outlook toward families or neighbors or people in general. One does not replace the other. But as people move in wider circles and deal with a more diverse range of people, the argument goes, they tilt in the direction of universalistic concepts.

A slight tilt of this kind is evident among farmers whose daily activities no longer revolve around a particular farm but involve multiple pieces of ground scattered over larger distances. The special attachment they may feel to the "home place" remains and yet is located within a more general concept of the various farms involved and the overall ways in which farming in the area is conducted.

Those differences do not appear as frequently, though, as farmers' own sense of a changing day-to-day relationship with the land. The slight change in sound that a tractor's engine makes when pulling through a patch of tough caliche is less evident when the tractor is larger and more powerful. The intuitive knowledge gained from running one's fingers through the soil is less important. A farmer squeezing a kernel of wheat would illustrate greater continuity with the past, while a moisture-monitoring device on a combine would illustrate greater discontinuity.

The farmers we spoke with described these changes mainly in terms of technological improvement and scientific measurement. Acreage that used to be measured by walking a field in yard-length steps was now measured by satellite imagery. An experienced farmer could still guess the approximate yield by walking into a field of wheat or cotton, but computerized information could now tell the exact yield for each part of the field. Seeds were distinguished by the numbers assigned to them by agribusiness companies rather than by feel.

There was no regret in farmers' descriptions of these changes, only an awareness that the meanings of farmland were shifting as a result. A farmer could still take pleasure in walking through a field, watching seedlings grow, and smelling new-mown hay. The real business of farming, though, involved managing information.

134

STEWARDSHIP OF THE LAND

"Farmers are always good stewards of the land." This is Mr. Rayburn, the cotton-belt farmer. "Always" sounds like he is deliberately overstating the case. He is. He knows of farmers in his community who use too much spray or who cheat a bit on irrigation restrictions. His point is only to emphasize that farmers have a strong stake in taking good care of their land.

The area where Mr. Rayburn farms used to be grassland. Prospectors considered it too dry for crops. Early settlers tried to grow wheat on the wide-open fields, but the land was too arid for wheat. They tried cotton and found it better able to withstand dry seasons. The water table was too low for tiling and ditch irrigation. Once rural electrification was completed, though, pumps could be installed. By the 1980s about 40 percent of the cotton was grown on irrigated land.

That was fine until it became evident that the underlying aquifer was being depleted. Now Mr. Rayburn and his neighbors have to abide by regulations restricting how much water they can pump each season. Although Mr. Rayburn sometimes chafes at the restrictions, he thinks they make sense and are in farmers' best interest. He considers being able to adapt to new ideas just what farmers do. His father, he says, "wasn't a stranger to trying new things." He sees himself doing that as well.

The water-use restrictions in Mr. Rayburn's area are a good example of how farmers think about land stewardship. Although they are as interested in monthly returns as anyone else, they understand the importance of thinking of land stewardship in the long term. The ones we spoke with were critical of corporations that think in the short term. They regarded mining companies that stripped the hillsides looking for coal as an abomination. They felt the same way about oil companies that turn whole fields into sludge ponds. They knew it was possible for farmers to throw on extra fertilizer and double-crop the soil until it was depleted. They considered it foolish to do this.

Thinking long-term is embedded in farmers' worldview. They talk of seven- to ten-year cycles. Seven-year cycles have biblical connotations. The Bible speaks of seven good years and seven lean years. Ten-year cycles are similarly inscribed with ups and downs. Several of the farmers we spoke with said they expected to lose money at least two or three times in a decade, make money two or three times, and break even the rest of the time.

The longer-term perspective stems partly from family history. Family land that a farmer's grandparents and parents cared for deserves to be farmed carefully as a kind of trust. Mr. Rayburn says the land he farms is his legacy. He likens farming to marking one's turf, not the way an alpha male territorial animal would, but not entirely different either. When farming is in the blood, the blood is somehow associated with the land.

Rented and newly acquired land deserves care for similar reasons. It was part of someone's family history, probably a neighbor's. It was Ben's land or Foster's land. Their spirit is still present. Whoever currently farms the land will be honored by those who follow because they took good care of it.

Thinking long-term seems especially to be related to farmers' musings about the land having been there for eons. This was especially evident in one of the farmers' comments about American farming still being an experiment only several hundred years old. That was not long at all, he thought, compared to the length of time the soil had been there and hopefully would remain. Another farmer referred to the agricultural history he had studied in college. It reminded him to anticipate cycles in agricultural production and to think long-term rather than focusing too much on seasonal variations.

We found this emphasis on long-term responsibility to be as true among people who did not expect their children to farm as among those who did. A woman who operated a dairy farm with her husband put it well. She admitted having a "jaded view" of farming and was not encouraging her children to farm. But she added, "I deeply appreciate our responsibility for the land."

Neighborly relations help define and monitor the meaning of good stewardship as well. Although these relations may be weaker and more distant than in the past, they still operate. Farmers' comments about success in acquiring land for rent or purchase illustrate the role of these relations. Weeds, poor crops, and soil erosion indicate a lack of attention to good stewardship. A landowner wants a farmer who will take better care of the land.

The definitions of good stewardship in farming communities are never crystal clear. As farming practices change and as new technologies come into use, ambiguities arise. The water-use restrictions in Mr. Rayburn's area are designed to ensure that an equivalent of at least 50 percent of the currently available underground water will still be available in fifty years. The 50-50 plan, though, was a matter of negotiation. It provided one definition of good stewardship. How it will work out depends considerably on unknown factors having to do with climate change.

Other uncertainties have arisen from no-till farming practices and new seed varieties and chemical innovations. The notion that farmers are *always* good stewards is pushback against critics who argue that farmers are ruining the land by misusing these innovations. No-till farming retains more of the moisture in the soil, farmers say, and it requires burning less fossil fuel than frequent cultivation would. But there are critics who contend against these claims. New seed varieties and pesticides are also difficult to assess in terms of long-range stewardship considerations.

If no-till farming and chemical innovations allow farmers to spend less time in close proximity to the land, sensitivity to the importance of good stewardship may be increasing in other ways. The wheat-belt farmer who has land he has not tilled for nearly twenty years says there has been a shift from thinking about the land "as just dirt" to recognizing that the soil is a "living fragile thing." The idea is that farmers' appreciation of the land is increasing as a result of advances in soil science and agronomy. Not decreasing. And that good stewardship necessitates understanding these advances. He attends a farming conference every winter to learn more about soil bacteria and microbes.[8]

Another wheat-belt farmer expressed a similar idea but put it somewhat differently. His appreciation of the land stemmed from working it with his father and grandfather, but also from earning a college degree in agronomy. "We're looking now at micro-nutrients—zinc, sulfur, magnesium," he explained. "All we used to worry about was nitrogen and phosphorus." The land he farms has been in wheat for a hundred years. "Every time you take a crop off, you mine something out of the ground. So we have to put more back in, just because of the nature of the land." Or as another farmer put it, "The ground is just like a horse. Sooner or later you have to feed the horse."[9]

Questions about good stewardship inevitably connect in farmers' minds with concerns about government regulation. Discussions of these concerns sometimes portray farmers as knee-jerk opponents of regulation. The farmers we spoke with included some who reinforced that image. But opinions were mixed. They included an understanding and appreciation of the need for regulation as well as concerns about its specific implications.

"I think some regulations are fine" was how a farmer who had been growing cotton for nearly half a century put it. He thought it was impossible to keep the land and food and fiber supply completely 100 percent free of contamination, but he considered it important to try. His concerns were mostly about city people who he felt did not understand what farming was all about. In his view, lawns and parks were far more likely to be treated with chemicals that contaminated the land than farm ground was.

"Walk into any Walmart and look at how darn many gallons of chemicals they have for sale there," a wheat farmer who was even more riled about criticisms from city people exclaimed. He thought farmers in his area were doing everything possible to avoid contaminating the soil. As farms were becoming larger, more of the farmers were hiring specialists in agronomy to make sure the right chemicals were being used. "The farmer knows better than anyone else about ruining the ground," he said, "because if he does, he damn sure (excuse my French) isn't going to have a living next year."

The specific regulations that drew greatest support were ones that protected the land and the health and well-being of the farming community itself. That included preserving the water supply, controlling noxious weeds, combating erosion, and avoiding chemical contamination of the soil. It did not extend to topics that farmers considered ridiculous, such as keeping dust particles from the air. There were examples of farmers participating voluntarily in soil conservation studies, reducing their use of chemicals and pesticides, and taking extra measures to ensure compliance.

If they were their own bosses, it was in their personal interest to comply with these protective measures. If they had employees, they might be more lax, but understood the possibility of fines and other penalties. As one farmer noted, it was sometimes difficult to keep farmhands out of fields that had recently been treated with chemicals, but it helped to do what his neighbor had done. He told the fieldworkers that their "balls are going to fall off" if they went into the fields too soon.

It was typically the paperwork that they disliked more than the regulations themselves. There seemed to be endless forms. They knew it was necessary to fill out the forms to prove that they were in compliance with the regulations. But lately it seemed that the number of forms had proliferated. Time spent doing paperwork was time not spent doing what they loved. It was time in the office rather than in the field.

"You have to make sure every T is crossed exactly right and every I is dotted exactly right and every word is in its place," a livestock and grain farmer complained. "Otherwise you're in trouble." He felt like he was spending far too much of his time at the Farm Service Agency (FSA) filling out water use reports, chemigation reports, chemical use reports, and fuel storage reports.

His wife took only a slightly more charitable view of the FSA. She agreed that the regulations took too much time. She worried that regulations seemed to lag behind the scientific and technological innovations that were helping agriculture advance. Besides that, there were deadlines to worry about and some of the regulations were unclear. "The poor little ladies in there at the FSA office have to take the farmers' anger," she said.

With added paperwork came the fear that small farmers would be unable to compete. If it became necessary to fight lawsuits brought by the government, then large corporate farms would be in a better position to hire lawyers, farmers said. To the extent that lobbyists influenced regulations, it also seemed likely that large corporations would shape things to their benefit rather than to the advantage of family farms.

"Big companies can have corporate lawyers on hand," a farmer in the corn belt observed, "but we can't hire a lawyer to have on staff." He was doing well, farming about twice as much land as his neighbors and raising hogs as well as corn and soybeans. That put him just at the level where he felt threatened by truly large corporate farms.

Hiring a lawyer was actually the least of his concerns. The regulations, he felt, were necessitating expansion in order to remain cost efficient. A farmer with a 500-gallon fuel tank, he feared, would have to spend $20,000 on a ring dike to satisfy the government's new safety regulations. Any standing water larger than a mud hole would have to be monitored and drained.

"It just eliminates competition," he says. "What scares me is Washington listening to lobbyists rather than farmers." He shakes his head. "Maybe there's some bad farmers, but I think in the long haul we take better care of the land than what they're trying to do with regulations."

Farmers' sense of being their own boss is clearly an important part of their thinking about regulations as well. The land might ultimately belong to God and it will be there long after they are gone. But in the meantime they consider it theirs. They own it or they rent it. And that makes them feel they know best how to take care of it.

"We don't like anybody telling us what we have to do," a cotton farmer admitted. "We like our freedom. We dislike people coming in and telling us what we have to do with our ground. Most of us are good stewards of the ground. That ground makes our living."

Whatever else it means to them, whether it has been in the family for generations or is a source of pride because they have earned the right to farm it, and whether it is a source of day-to-day pleasure or frustration, the fact that land is an asset is never far from farmers' minds. They feel fortunate if they bought land that has risen in value. When annual net returns are low, they can mostly look back at steadily rising land prices and hope those increases will continue.

The old saying that "they're not making any more of it" was one we heard repeatedly in our interviews. The limited supply of land relative to a growing population implied that land values would steadily increase over the long haul. Farmers with small and large holdings alike talked about drawing income from the land when they retired and passing it on to their children as a legacy.

"We have some friends who don't farm anymore and they have inherited a little bit of land," Mr. Sellers observes after talking at length about his own land and how land values in his area have risen to record levels in the last few years. "I tell them, 'The dumbest thing you could ever do is sell it. Set it up in a trust and when you die it goes to your kids. Every year they get a check from it and they can take a trip or pay the kids' college tuition.' Some people sell it. They sell it, it's gone. I think it's crazy to sell it."

Farmers' relationship to the land is in the final analysis a delicate balancing act. It involves an intimate emotional attachment similar to parents' relationship with children. The attachment stems from familiarity and from a significant investment of time and energy. It is deepened by the family ties that are often woven into the land and by the stories that turn the land into a living legacy. The relationship is equally a matter of dollars and cents.

In this respect considerations of intimacy and money are never far apart. It can seem awkward to think of land that is so personally meaningful as having a monetary value, just as it can putting a price on a child for purposes of a life insurance policy. Hanging onto the farm for sentimental reasons can overpower questions about rational return on investment strategies.

Holding the aesthetic and the economic aspects of the land in tension nevertheless serves a positive purpose. The aesthetic part makes farming more enjoyable through good times and bad than if it were simply a job. The aesthetic and economic parts come together in thinking about sustainability and appropriate levels of regulation. Farmers describe good stewardship in terms of taking both into account.

TECHNOLOGY

6

With all the new technology that's coming out, you have to decide what to spend money on. In my grandfather's day, things weren't changing that much. It's a lot more dynamic now to stay competitive.

—Wheat-belt farmer, male, age 32

He'd be out there on the tractor spraying. He ended up getting so sick that he couldn't even eat. We finally realized it was the spray.

—Corn-belt farmer, female, age 69

One way of describing the long-term changes that have taken place in American society is to say that the center of gravity has shifted from farms to cities. That version suggests that agriculture was important in the seventeenth and eighteenth centuries but by the end of the nineteenth century industrialization had become the dominant influence. Then during the twentieth century agriculture fell further behind as technology created a larger industrial, commercial, and service economy.

The missing piece in that version is the fact that technology has also greatly influenced how farming is done. These influences have included the shift from human and animal-based labor to work performed by engines and machines. Mechanized labor itself has changed as machines have become larger, more powerful, and more technologically advanced. In recent decades technology also includes greater use of chemical fertilizers, pesticides, and genetically engineered seed varieties.

Since the 1980s farming has been profoundly influenced by the information technology revolution. Although these innovations are still unfolding, their impact is already as significant as previous waves of innovation such as the transition from horse-drawn to engine-powered equipment. Tractors, combines, and cotton pickers are not only larger. They are also equipped with computerized monitors and satellite guidance systems. Information technology has not only brought cell phones and home computers to farms. It is also the source of new chemicals, pesticides, and genetically engineered seed.

Many of these technological innovations have sparked controversy. The possible effects of chemicals and genetic engineering on the safety and quality of the nation's food supply have been particularly controversial. Questions have also been raised about the effects of technological innovation on the soil, air and water, food prices, and food supplies.

Although bureaus and leaders other than farmers themselves largely determine the laws and policies regulating technological innovations, farmers' daily lives are shaped by these innovations. Staying in business involves decisions about purchasing newer and better equipment and using the latest chemicals and seed varieties.

These decisions are matters of bottom-line financial considerations but also of emotion, sentiment, and values. How farmers think about technology reflects their understandings of themselves, their families, their work, and their relationships with the land. These understandings are in turn affected by the advent of new technologies and their use.

Who adopts and who does not adopt technological innovations are perennial questions among agricultural economists. Studies identify a variety of important factors, such as the size and location of farms, the age and education of farmers, and access to credit and other sources of financial capital. The relevant factors can be conceived of as constraints or as facilitating conditions.[1]

All of the farmers we spoke with had adopted new technologies. The question was not whether they had or had not. It was rather what technological innovation meant to them—how they felt it benefited them, not only financially, what its costs were, and how it was affecting their relations with their families and the land. These were topics they wanted to discuss.

In our interviews we asked farmers to talk about decisions they had made about investing in newer and larger equipment. We asked their views about genetically engineered seed and new ways of controlling weeds and insects with pesticides. We were interested in farmers' perceptions, not only of whether these innovations were beneficial, but also of the pluses and minuses and how farming was changing as a result.

The responses demonstrate clearly that technological change and adapting to these changes are among the most important decisions with which farmers are concerned. Their arguments naturally reflect differences in location and scale. Differences notwithstanding, technology also represents a measure by which farmers evaluate their own status as farmers and position in their communities.[2]

Adopting new technology is driven by the perceived necessity of staying competitive by achieving greater efficiency and higher yields. It is also driven by farmers' desire to be on the cutting edge. Experimenting, innovating, and being the first to try something new is of perceived value in itself.

The ambivalence toward technological innovation that frequently emerges in farmers' comments registers awareness that technology involves high costs and risks as well as potential rewards. Rarely does the ambivalence suggest that farmers are wedded to old-fashioned ideas or are reluctant to change.

Their ambivalence, though, poses the question of whether the current information technology revolution is helping to preserve family farming or is leading toward its demise. Farmers are unsure of the answer. But they worry. Is the convenience of better machinery and is the appeal of higher yields worth it? Or are these developments putting even the best farmers in jeopardy of being subjected to or replaced by large agribusiness conglomerates?

These are questions central to modern culture itself, not just to farming. Whether the technology in question was steam-powered engines or the automobile, the quandary was how to adapt without sacrificing what had been considered most valuable to family lifestyles and communities. The persons directly involved have had to engage in cultural work to make sense of the decisions they were making. Farmers' ways of thinking and talking about these decisions are an illuminating window into this important cultural work.

BIG MACHINERY

Nearly all the farmers we spoke with identified larger and more powerful equipment as the single most important change they had experienced during their lifetime and in comparison with the previous generation of farmers. Tractors, cultivators, planters, sprayers, and combines were all capable of covering significantly more ground per day than earlier models. Although the cost was much higher, farmers consistently felt that the new equipment was worth it.

Examples of newer and larger equipment can be found at almost any decent-sized farm. A visit to a large cotton farm is likely to reveal an eight-row self-propelled cotton stripper and boll buggy. Possibly two of them. A corn and soybean farm of comparable acreage is likely to include a 40-foot, sixteen-row corn harvester.

"We've got two 40-foot soybean headers," a corn-belt farmer we spoke with noted. That compared with 15-foot headers a few years ago. He used to have a six-row planter. Now he had a thirty-six-row planter. He could plant as many acres in an hour as his father did in a day. Another corn-belt farmer showed us a new $200,000 sprayer he had recently purchased. It had a 120-foot boom. The boom included a sonar device that kept each section positioned at exactly the same height above the crop.

The shift toward larger machinery in wheat farming was evident at Mr. Bower's place. He owns two combines. One cuts a 25-foot swathe each time it goes through the field. The other cuts a 30-foot swathe. These machines have replaced the 14-foot combine he harvested with a few years ago.

During harvest Mr. Bower drives one of the combines and his brother or a friend in the area who does not farm drives the other. Mrs. Bower drives a tractor pulling a grain cart. The combines empty wheat without stopping into the grain cart, which then unloads the wheat into an eighteen-wheel truck for transport to the grain elevator in town. Harvest that used to take eight days and run the risk of hail and rain now takes half the time.

Other farmers we talked with praised the newer combines for being not only bigger and faster but also better. They liked the on-board monitoring system that told them exactly how many acres they had cut, what the yield was for each acre, and what the moisture content was. With different seed planted to maximize the yield in different parts of the field, they could tell whether they had made the right decision or needed to make adjustments next year. Several of the farmers had tractors with on-board computers that recognized which implement was being pulled and provided relevant information about how much seed was being planted or how much spray was being applied.

Real-time moisture monitoring had another side benefit, one farmer suggested, half seriously. Suppose you started harvesting soybeans a bit too early when the plants were still damp and you found the moisture content was one percent too high. You could go ahead and harvest a couple of bin loads and put those in the bottom of the truck. Cover them with drier bin loads cut later. The tester at the grain elevator would never know.

The language farmers use to describe machinery is an indication of how they value it. The language typically emphasizes the increase in acreage per hour or day that newer and larger equipment provides. "We can do 100 to 140 acres a day harvesting hay," a farmer who recently purchased a new mower and windrow merger implement observed. "We never used to be able to do that. We were lucky to get thirty with our other equipment."

A corn and soybean farmer with about four thousand mostly rented acres remarked, "I started with an old open tractor and a four-row planter and now I've got two thirty-one-row planters." "If we cut six hundred bushels a day," another corn-belt farmer noted, "we used to think that was a good day; now we cut fifteen thousand bushels a day."

Most of the farmers we spoke with said bigger and more powerful machinery was not simply more efficient. It was fun. They took pleasure in purchasing one of the largest tractors or combines on the market. "Maybe it's just a childhood thing," a third-generation wheat and livestock farmer notes with some embarrassment, "but I just enjoy operating big machinery." "It's just fantastic," the corn and soybean farmer says. "It's a great pleasure to go crawl in a $250,000 tractor and go to the field for the day or crawl in a $500,000 combine for the day. It's a pleasure. It really is."[3]

Other farmers we spoke with acknowledged that purchasing newer and bigger machinery was like driving a new car. It might get better mileage and need repairs less often. But it also made you feel good. It was nice to experiment with the latest gadgetry. The neighbors might not say anything about it, but they knew your machinery was bigger and better than theirs.

Apart from the sheer pleasure of it, the newer machinery significantly reduced the physical energy required. Sitting in an air-conditioned tractor cab all day was far less exhausting than driving a tractor with no cab. "It's a lot easier to run all day and all night," the same farmer explained. "You feel human at the end of the day instead of getting the snot beat out of you all day long," the operator of a large dairy farm remarked. Another farmer who had been in the business for nearly half a century observed that even four-wheel drive made a significant difference. It was much easier to turn around at the end of the field, he said, than having to jam hard on the brake to turn a two-wheel-drive tractor.

The major advantage of larger and more powerful machinery was being able to farm more land. That made sense for large farmers and for farmers who could rent or purchase more land. It was less attractive to farmers with smaller acreage. A farmer with two hundred acres of land he owned and another two hundred he rented explained it this way: "My grandfather farmed this land with horses. In the 1930s he bought a tractor for $900. In the 1970s we paid $9,000 for a tractor. Today you pay $90,000. Something is off track here."

There was also an undercurrent in farmers' comments suggesting that more expensive equipment was as constraining as it was liberating. The constraint involved having to work harder and longer or acquire more land in order to pay for the equipment. Some of the farmers who otherwise favored larger machinery and had invested in it nevertheless voiced doubts similar to those of critics of other kinds of industrialization. Just as an assembly line shaped the lives of workers in not always desirable ways, so mechanization influenced the meaning of farming.

One of the more successful farmers we spoke with put it this way. "You could just work a horse so many hours a day." It had to be fed and rested, which meant that farmers' work days were limited as well. But mechanization changed that. "They put lights on tractors. They should have never been allowed. You could run a tractor twenty-four hours a day. The real aggressive people, the workaholics, they would work till midnight and then get up and start again at three."

That was his example of how farming had changed in previous generations. The recent changes were equally worrisome, he felt. "It's kind of a catch-22," he said. A new air-conditioned tractor or combine offers comfort compared to the old ones. "But the new equipment—a $130,000 tractor or a $400,000 combine—is not easy to pay for. You buy them, get them financed, and then feel forced to farm more ground to pay for it."

"It looks big," he added. "It's nice equipment. But the truth of the matter is that farmers' lifestyle and happiness within the family is not like it used to be. It's really become different in the last twenty or thirty years. It's just run, run, run."

"Technology is more of a burden than a benefit," a dairy farmer suggested in talking about the changes he has experienced over the past three decades. The point of laborsaving equipment such as pipeline milkers and milking parlors and huge forage cutters, he said, was to give farmers more time for other things. "Bullshit! You've got to work more acres, milk more cows, and grow bigger. I can't take Sundays off. It isn't any easier. It's just more work."

Another farmer offered a critical perspective on machinery as well. Although he was quite successful and farmed enough land to make large machinery necessary, he was not keen on having to invest in newer and larger equipment. "I hate machinery!" he said. "It's nothing but a sink for expenses." He recently purchased a larger forage chopper and a newer corn planter. But he says he only did so because it seemed necessary in order to remain efficient. "It takes a lot of time and a lot more labor to keep them operating."

Comments from farmers who are less than enthusiastic about newer and larger machinery serve as a reminder of how closely the evolution of farming over the past century has been related to advances in mechanization. While earlier generations of farmers developed skills in working with horses and other farm animals, the interest and ability to work with machinery has been far more important for many farmers in recent generations.

Farmers' thoughts about machinery indicate the importance of having mechanical skills and training, being able to repair and maintain machinery, and indeed finding pleasure in operating it. The reservations a few farmers express about machinery suggest that farming has perhaps been shaped by mechanization even more than farmers may have wished.

The question this relationship with machinery poses for the future is whether larger and more specialized equipment will fundamentally alter the family aspects of family farming. Farmers we spoke with agreed that rising equipment costs were affecting the financial aspects of family farming. There was less agreement about the effects on family relationships.

The negative effects farmers described included machinery becoming too sophisticated for children to operate the way they might have in earlier days. Parents thought it was harder to transmit practical skills. Some regretted not being as available from day to day. Instead of working near home and regularly seeing their children, they were off at a distance operating a large machine.

There were plenty of examples, though, of farm families adapting to larger and more specialized machinery in ways that maintained close family relationships. Even though the equipment was more expensive, some aspects of it were easier to operate than in the past. Some of the wives drove equipment they said would have been too difficult a generation ago. Some of the farmers still active in their seventies and eighties said the same thing. There was still a division of labor in which spouses and children could do their part.

These family relationships were clearly evident at the wheat and soybean farm Elizabeth and Willard Armstrong operate with their three teenage children. Mrs. Armstrong had just come in from taking a load of seed wheat to the field when we caught up with her, and she was getting ready to drive her daughter to town for a piano lesson. She described how larger equipment has necessitated the whole family working together.

The Armstrongs' busiest time is harvesting wheat. They used to harvest with two smaller combines that were older and frequently broke down. It took two people to operate them and sometimes a third to run into town for spare parts. A few years ago they traded the older combines for a new one. It cuts a 35-foot swathe and rarely breaks down. A typical harvest day begins at ten o'clock when the overnight moisture has dried and ends twelve hours later.

Each family member has a job. Before the fieldwork begins, they pack sandwiches, fill water jugs, and collect gasoline, oil, and grease. The idea is to work straight through without stopping for lunch or supper. The two girls prepare the food. The son services the machinery. Mrs. Armstrong records the wheat tickets from the grain elevator for the previous day and tallies the amounts for each of their landlords.

The fields are all within seven miles of each other. The 35-foot header is too wide to take on the roads. One of the daughters is in charge of the header trailer. The other daughter and son take turns driving the tractor pulling the grain cart that carries the wheat from the combine to the truck at the edge of the field. Mrs. Armstrong moves an empty truck into position and drives the full truck to the elevator in town. Mr. Armstrong operates the combine. The quiet air-conditioned cab makes it possible for him to explain the procedures to one of the children who rides with him.

The Armstrongs involve the children because they consider it important for children to learn from their parents. This was one of the reasons they decided to farm in the first place. "The main thing is that they can see what their parents do," Mrs. Armstrong says. "Even though my kids aren't involved the same way my husband was when he was a kid, they're still involved." Technologically sophisticated equipment has not changed that, she says. "They know what it is to be responsible. They know what it is to be needed."

GLOBAL POSITIONING

In addition to larger and more expensive equipment, nearly every farmer we spoke with mentioned global positioning systems (GPS) as the technological innovation that was revolutionizing farming. Instead of walking the fields to be sure how many acres they were planting, they relied on satellite images. Their reports from the county Farm Service Agency office included satellite images showing exactly what was planted in each field. Most of all they appreciated GPS guidance mechanisms on their equipment.

"I took my dad out after our son got the new tractor," one of the women we spoke with said. "Dad, you've got to see this," she told him. "Dad, our son did not touch the steering wheel the whole time he was driving the tractor just now!" Her father could hardly believe it. "No, it's on GPS and it knows right where to go." She added that her son absolutely loves the new technology.

Her son acknowledged that the GPS-guided tractor was a pleasure to operate. The benefit was not only that he was less tired after a long day on the tractor. The main benefit was that he could get the cotton planted at just the right time, rather than having to worry as much about being delayed because of too much or too little rain. That was partly because he was operating a sixteen-row planter instead of a four- or an eight-row planter. But it was mostly because he and a hired man could plant for eight or nine days around the clock. With GPS they could plant all night. They did not have to see the field to plant it.

An elderly couple in corn and soybean country was similarly impressed with one of the fields a renter of theirs farmed. "You could have taken a rifle and shot down between the rows and you wouldn't have touched anything on either side," the older man recalled, "just nice and straight all the way as far as you could see." He complimented the renter. The renter replied, "I can't take much credit. All I do is lift the drill at the end of the field, turn around, put the drill back down, and press a button."

On Mr. Bower's wheat farm, the principal advantage of GPS guidance has been enabling him to experiment more efficiently with no-till farming. A decade ago, he followed the traditional pattern of plowing fields shortly after harvest and then cultivating several times to prevent weeds before planting in September. Plowing made a furrow that showed clearly where the field had been plowed and where to drive the tractor. But no-till farming involves no plowing or cultivating between harvests and planting.

Fertilizer is applied and spraying controls weeds. Driving a tractor through an uncultivated field or a self-propelled sprayer with a 90-foot boom makes it difficult to be sure of the line between the sprayed and unsprayed part of the field. GPS guidance automatically positions the tractor or sprayer at the right place.

Besides those advantages, another wheat-belt farmer noted, GPS was beneficial even for ordinary harrowing or planting. Where he lived, the fields were so large and trees were so scarce that wind-blown dust was common on all but the quietest days. He recalled times on the tractor before GPS when he literally could not see where he was going. GPS has solved that problem.

A farmer in corn-belt country illustrated several additional advantages of GPS technology. During planting season he sometimes begins at 5:30 in the morning and continues until 2:30 or 3:00 the next morning. Driving the tractor that many hours used to leave him with sore shoulder muscles for a week. Now, he says, it's "no big deal." Not to mention being able to plant at night. With GPS mapping, he and the farm service agent work out exactly how much he is planting and how much he is setting aside for conservation. When crops are harvested, he immediately knows the yield for each field—information he takes to the banker for next year's loan. "The technology is just unbelievable," he says.

Another corn-belt farmer emphasized the role GPS mapping plays in his planting decisions. He employs what he calls variable rating. In fact, he says, "we variable rate now," turning the phrase into a verb. "Variable rating" means planting more seed in some parts of a field than in others. The GPS map includes a layer describing the soil quality within each grid. As the tractor moves from grid to grid, the monitoring system automatically varies the amount of seed the planting equipment sows.

"Our combine has all the GPS recording equipment," a wheat-belt farmer who harvests about a thousand acres annually remarked. He takes the information from the combine's computer, downloads it into this office computer, adds soil sample data, and then loads it into the computers on his sprayer and fertilizer applicator. The information tells where in the field to apply different amounts of fertilizer.

The cost of GPS put it out of reach for many of the smaller farmers we spoke with. At ten to fifteen thousand dollars plus a monthly operating fee, it was a luxury they felt they could do without. There were other ways to accomplish some of the same things. One was a laser system that showed where to spray or plant. An even cheaper method was a kind of nontoxic dye that marked the line of sprayed and unsprayed or planted and unplanted.

But GPS had become so popular that the farmers we spoke with who could not afford it admitted it was something they wanted. One farmer went so far as to say that he "lusted" after it. They wished it was economical and sometimes consoled themselves that GPS was more about status and feeling that a person was successful than anything else. "Like I've told people," a farmer who felt this way explained, "my farming is to make a living, not to make an impression."

The farmers who had invested in GPS were confident, though, that the investment was worth it. The better systems could guide a tractor within two or three inches of accuracy. With seed and fertilizer costing as much as they did, that kind of accuracy eliminated duplication and saved thousands of dollars.

The other benefit of GPS was better coordination of seed and fertilizer with specific variations in soil chemistry and composition. GPS facilitated the process of mapping fields into grids, taking soil samples from those grids, and adjusting seed varieties and fertilizer accordingly. Comparisons of yields per grid provided information for making further adjustments the following season.

INFORMATION TECHNOLOGY

Besides its role in global positioning, information technology of other kinds was having a significant impact on farmers' activities. The most obvious benefits mentioned by the farmers we spoke with were from cell phones and the Internet. In earlier decades farmers were out of touch while doing fieldwork unless they stopped and walked or drove home to use the telephone. That was less of a problem when most of the land was near the house. It was more of a problem when the land was scattered. As an older woman we spoke with recalled, "I would have to drive to three different places to find my husband if something came up." Two-way radios and walkie-talkie connections helped for a while but were still limited. With cell phones, farmers could call home from the tractor or combine.

Cell-phone technology had rather different implications for farmers depending on the kind of farming they did and their scale of operations. Smaller farmers and farmers who farmed alone embraced cell phones as a hedge against isolation. Getting stuck in the mud or having an equipment breakdown miles from home no longer constituted the difficulty it once did. A call home or to a neighbor could bring help. Farmers with large-scale operations tagged cell phones as the key to better coordination. It was now easier to manage employees at a distance. "Being spread out has become much easier," one farmer explained. "We can talk to each tractor whenever we have to."

Nobody thought cell phones would be a decisive factor in farms becoming even larger in the future. Other factors having to do with capital, equipment, and available land were considered more important. But if farms did become larger and if farming shifted to involve more hired employees, cell phones would help.

The largest effect of information technology was in connecting farm families more quickly and efficiently with relevant sources of information. The benefits mentioned most frequently included online links to grain and futures markets, opportunities to purchase supplies and equipment online, and connections with agronomists, county agents, and agricultural departments. Information technology provided wider opportunities for keeping up with general news and knowledge as well.

Most of the farmers we spoke with used home computers to keep track of expenses and revenue, to file tax reports, and to fill out government forms. As in other lines of work computerized information enabled possibilities for more detailed record keeping. Interesting examples were given by several of the dairy farmers we interviewed. As cows filed through the dairy parlor for milking, the milking system provided a computer-generated record of how much milk each cow produced. Attention could then be given to any particular cows whose output indicated a problem.

An unexpected consequence of information technology, several farmers reported, was that government regulations were becoming more flexible. There was still some concern that government bureaucrats were collecting and storing more and more information, whether in the form of crop maps, tax reports, or mandated farm censuses. But computerized information was at least enabling government programs to be tailored to individual variations. For example, the farmers in one community had been able to work out a cafeteria plan for meeting standards regulating atrazine runoff. The plan took account of differences in crops, tillage, and rainfall and provided several options for meeting the standards.

The consequences of information technology for social relationships in farming communities are more obvious. Some of the consequences are beneficial. Cell phones make it easier for farmers to communicate. The Internet facilitates information about school activities, farmers' co-op meetings, and other local events. Computer links enable a single central office to handle grain elevators at a dozen scattered locations. In several of the smaller communities we visited the nearest equipment dealer was thirty or forty miles away. Parts could at least be ordered online.[4]

The farmers we spoke with appreciated these benefits, but they worried about the negative effects on their communities. As we saw in chapter 2 buying and selling via the Internet was making it harder for local businesses to compete. A farmer who bought and sold cattle on the Internet, for example, said he was pleased to have access to wider markets, but he regretted that the local auction barn was no longer in business. Another farmer was finding it cheaper to purchase fertilizers and chemicals through the Internet, but he figured it was contributing to the demise of his rural community.

We heard some complaints about the companies providing Internet service, too. Just as there had been in previous eras with telephone and television service, there were complaints about outages and poor reception in rural areas. It seemed to some of the farmers we spoke with that big powerful companies were taking advantage of farming communities.

The older farmer who complimented his renter on straight rows illustrated one of the concerns about big companies. A guy drove into his yard one day, he recalled, and offered to pay a hundred dollars a year for ten years to put up a cell-phone tower on a 10-foot by 10-foot cement slab in the farmer's pasture. The contract included no provisions for removing the slab if the tower ceased to be used. It did not preclude additions to the tower. After consulting his attorney, the farmer decided against signing the contract. The incident had taught him a lesson. Information technology might be the wave of the future, but it could be an invitation to trouble as well.

Whether information technology was altering the social structure of farming itself was harder to determine. On the one hand, cell phones and home computers were inexpensive enough that farmers large and small were joining the information revolution. On the other hand, some of the new technologies were out of reach for all but the largest farmers. Those were also the farmers best positioned to participate in information-driven developments in the wider economy.

Fred Young, a third-generation truck farmer who has been in business with his brother since the late 1960s, gave one of the most striking examples of farming adapting to these wider developments. Through the 1980s they grew apples, pears, and cherries at a single location. In the 1990s they expanded. They now have orchards at seven locations. They annually ship more than a hundred thousand bins of apples.[5]

Information technology has played two important roles in Young Farms' expansion. The first has involved producing and keeping track of significantly more diverse consumer-driven output. In the 1980s the Young brothers grew Red Delicious and Golden Delicious apples and only a few cherries. They currently grow nine varieties of apples and seven kinds of cherries. The newer varieties of apples include Braeburns, Fujis, Granny Smith, Honey Crisp, Jazz, Pacific Rose, and Pink Ladies. Which trees to plant and where to ship the apples are based on computerized information. In addition computerized equipment has made it possible for Young Farms to develop a presized line, which means taking orders from wholesalers and retailers for boxes of apples sorted by size.

The other role information technology has played at Young Farms involves responding more quickly to orders. Large supermarket franchises such as Walmart, Sam's Club, and Safeway reduce overhead by keeping warehouse inventory at a minimum. That means they want orders filled almost immediately. "It's not uncommon for us to get an order at eleven in the morning," Mr. Young says, "that has to be packed and shipped by five that afternoon." The "daily fight," as he calls it, depends on having his own information available on a minute-by-minute basis about what the orders are, where his inventory is located, and how to get it shipped. "We used to sell our fruit because we had the best quality," he explains, "but now quality is just a given and we get the sales because we have the best service."

GENETIC ENGINEERING

Genetically modified organisms—or GMOs as they are popularly known—have been one of the most influential technological developments in recent years. Between 2000 and 2012, the proportion of corn planted in the United States that was genetically engineered increased from 25 percent to 88 percent, the percentage for soybeans climbed from 54 percent to 93 percent, and the proportion of cotton rose from 61 percent to 94 percent. The technology focused especially on insect-resistant and herbicide-resistant modifications.[6]

Genetic enhancement has been one of the most controversial topics as far as consumers and consumer watchdog groups are concerned. Among the worries are fears that humanly engineered genetic modification will produce illness in human consumers, lead to mutations in plants and animals that cannot be controlled, or result in patterns of breeding and modification that become vulnerable to environmental dangers such as weeds, insects, and parasites.[7]

The corn and soybean farmers we asked to tell us their opinions about genetic enhancement rarely expressed serious concerns, although they noted both advantages and disadvantages. Many of them said they had been planting GMO corn or soybeans for at least four or five years and had experienced no significant problems. "I don't perceive it as being that different from hybrids," one farmer explained, noting that cross-fertilization had been practiced for centuries. The only difference, he thought, was that genetic modification was quicker and more efficient because of knowing which particular "chunk of gene code" was being modified.

The advantage of genetic modification was that yields had increased dramatically. In addition production costs were often lower because less cultivation was necessary to control weeds. That meant less time in the fields, less wear and tear on equipment, and lower expenses for fuel. This was a point farmers old enough to have done things the hard way especially appreciated. "It's not great having to hoe," a cotton farmer in his late sixties noted, explaining what he liked about Roundup Ready cotton. "You go out there now, you spray it one day, you go home, and you don't worry about the weeds growing if it rains and they're not hoed."

Besides its effect on production costs, farmers we spoke with frequently argued that the commodities produced were of higher quality. The genetically modified cotton Mr. Rayburn produced, he observed, was simply better because the seed was pest and insect resistant. "I'm sending a better product down the supply chain," he said. Before, when cotton had worm infestation, the plants still produced, but the cotton was of lower quality. Sometimes it was stained. "The end consumer would like to see a quality product," he believes.

That view was shared by other farmers who saw GMOs as a more effective way to deal with insects than pesticides. In their view GMOs were more environmentally friendly than pesticides. Several of the truck farmers we spoke with thought this was true for fruits and vegetables as much as for small-grain crops and cotton. They noted that GMOs were making it possible to grow plants with less water in some cases and in others to time vegetables such as peas and beans so that they would ripen at different times and thus be available as fresh produce to consumers over longer periods. They realized that the public was not entirely on board with GMOs, though. As one farmer observed, vegetable processing companies would "get crucified" if they went too far in the direction of pest-resistant GMOs.

Farmers' enthusiasm for GMOs was heightened by expectations about the future as well as by recent developments. Genetically engineered seed corn that was temperature controlled to prevent it from germinating until the weather was warm, for example, fueled hope that corn could eventually be planted in the fall right behind the combine, rather than in the spring. Another hope was that seed would no longer need to be planted annually. It could instead be planted once and programmed to come up each spring on its own over a decade or so. Yet another expectation was that genetic engineering would improve the water use efficiency of plants to the point that less irrigation would be required or that different crops could be grown in arid locations.

The main disadvantage of weed-resistant GMOs and new herbicides was that weeds were adapting. These "super weeds," as they were called, were capable of surviving applications of Round Up and other herbicides. Many of the farmers we spoke with had seen this happen in their communities. They differed, though, in views about why it was happening. Most had enough experience with weeds to believe that weeds were capable of evolving and outsmarting almost any attempt to control them.

As a wheat-belt farmer whose conservative religious beliefs made it difficult to use the word explained, "evolution was going on all the time" among the weeds in his fields. "Some individual weeds are going to tolerate the herbicide better than most and, daggummit, if they survive and make seed, then we've got a bunch of those little critters that are tougher."

It was just a matter of keeping one step ahead of them, either by developing new herbicides and seed or by mixing cocktails of currently available herbicides. "Hopefully those chemical companies will put their nose to the grindstone and stay ahead of things" was how one farmer put it. Few considered it possible to reverse the process by going back to earlier practices. In for a penny, in for a pound, it seemed. Some blamed farmers themselves. As one observed, "Farmers try to save a buck" by using a weaker mixture of chemicals than recommended. Or as another admitted, he had made some mistakes with chemicals by focusing too much on the short-run bottom line.

Another disadvantage was that the price of seed has gone up dramatically. When Mr. Rayburn began farming in the 1980s, cottonseed was $20 a bag. He recently bought some that cost $350 a bag. He remembers when seed was hardly much of an input cost at all. It currently accounts for a quarter of his total expenses.

The steep hike in seed prices was evident in other areas as well. A corn-belt farmer, for example, estimated that seed increased from six dollars an acre to eighty or eighty-five dollars in recent decades. Last year he planted 240 acres of corn. The seed cost $24,000. Another corn-belt farmer said the cost in his community had jumped to $130 an acre. He was spending half a million dollars annually on seed.

Yet another disadvantage was pushback from consumers. This was more of a concern to dairy and livestock farmers than to cotton and small-grain farmers. For dairy farmers the concern focused on BST (bovine somatotropin), a natural peptide hormone produced by cows, but one that was also produced through recombinant DNA technology as an artificial growth hormone to increase milk production in lactating cows. The dairy farmers we interviewed felt it was best to avoid artificial BST even though they held mixed views about its actual health risks.[8]

Several of the farmers we spoke with were also concerned that genetic modification might be having ill effects on the land. They were happy that new seed varieties were producing higher yields and pleased that weeds were more easily controlled. But the direct relationship they used to have with the land was becoming less direct. It was now a relationship with chemicals. They used to know the land almost as if it was speaking to them, but now its voice was unclear. Perhaps the chemicals were okay, but it was harder to be sure.

"I heard a man talking on the radio the other night," a woman who had lived and worked on a farm all of her married life reported, "and he was saying all these chemicals are destroying our soil." The argument, as she understood it, was that nutrients were being taken from the soil but the plants being harvested did not yield as much nutrition, which in turn meant less nutrition for animals and less for humans. She was unsure what to think. Maybe the man on the radio was "just one of these crackpots." But she wanted more information. "Eventually we'll be destroying the earth with these chemicals," she feared.

The cotton farmer who appreciated not having to hoe felt the same way. It used to be you planted cotton and if it rained at the right time the plants almost jumped out of the ground and just kept on growing. Now it was harder to tell. Sometimes the seed just sat there. It was easy to make mistakes, too. A neighbor of his got mixed up and sprayed the wrong field and killed the entire crop. "We're getting into that deal," he exclaimed, "where you've got to keep your head out of your tail or you'll be in trouble!"

Apart from the economic costs and benefits involved, genetic modification and other new developments in chemicals and pesticides were affecting social relationships in farm communities. Farmers were unsure who to trust for reliable information when the information was increasingly technical and often changed from week to week. Should they trust their neighbors' opinion? Or did neighbors only tell what worked and keep mum about what did not? Did experts from agricultural extension services know? Were representatives from the chemical companies trustworthy?

Mr. Engstrom exhibited the frustration many of the farmers we spoke with felt in trying to address these concerns. He thought the new products were terrific. Genetically modified seed corn dramatically increased his yields. Weeding was no longer needed. Insecticides had their place. "You get an apple and it's got a big old bug in it, that isn't very good." The problem, though, was dealing with conflicting advice. Just the other day, he'd gone to a meeting and learned that a new fertilizer on the market should be put on early. But that same afternoon he went to another meeting and the speaker said, "If you put it on early, it will stunt the crop."

The difficulty for Mr. Engstrom was not that experts might have differing opinions. It was wondering if the experts were being objective. He worried that they were representing the big chemical companies' interests more than his. It used to be you knew the people who ran the local co-op, he said. But then the salespeople and managers representing the big chemical companies came to town and tried to change everything. "They all turned out to be assholes!"

Skepticism toward experts and chemical companies fueled farmers' sense that the independence they value so highly is being lost. That fear of losing control, as we saw in the last chapter, relates to government regulations as well. The concern here was not simply with paperwork or feeling that regulations favored large corporations instead of small farmers. It was rather that technological innovations were putting the knowledge needed to farm beyond the reach of the average farmer. If that was happening, it made it harder for farmers to know whether the regulations were justified or not.

"A farmer is raising kids out there on the farm," a corn and soybean farmer argued. "He's sure not going to go out there and put chemicals down to control the weeds that are going to make his kids sick." He felt the same way about antibiotics that would make his livestock sick. Or regulations about food and water. "They drink the same water that everybody else does. They eat the same food."

He was confident that new chemicals and seed varieties went through "years and years of testing" before they were approved for use on farms. And yet his arguments were based on trust in knowledge and testing that he did not understand personally.

Government was becoming the farmer's boss, in this perception, dictating what kinds of chemicals could be used, when they could be applied, and how water and air could be affected. They may not have liked the government looking over their shoulder, but the fact that it did was a basis for thinking that the new chemicals were okay. The reason not to worry about genetically modified seed or new pesticides, they said, was that all of these developments were carefully tested before being approved for farmers' use.

"As far as whether it's safe or not," a wheat-belt farmer in his early forties who was harvesting several thousand acres of chemically treated wheat observed, "I guess we assume and trust that the government will regulate all that." He was also realistic. "Will they know for sure that it's 100 percent safe or not?" he added. "Probably not for another fifty or a hundred years."

"There's not too much regulation," Neil Jorgensen declared. He said farmers in his community strictly adhered to a list of regulations involving chemical and fertilizer runoff, for example, and that those regulations prevented excessive nitrate levels in surrounding streams and water supplies. Like other farmers, he thought unrelated uses of lawn fertilizer and weed killers in cities constituted a more serious environmental problem.

When it came to the food they consumed, farmers were characteristically supportive of strict controls. They occasionally expressed concern about meat and poultry with additives or vegetables and fruit contaminated with pesticides. Like everyone else, they read the reports and heard about food safety concerns on television. They were unsure if what they were eating was safe. But the faith they expressed toward their own farm output generally included the farm products they consumed. They believed domestic products subject to US regulations were best.

"A lot of our food comes from overseas anymore where it doesn't have to be regulated." This was Mr. Rayburn speaking. "They don't know what's put on it. I have a bigger concern about what's coming across the border and going into my food chain than what I have locally." He was thinking especially about food coming from Mexico. "Lots of chemicals are banned here in the United States that are still in use there," he said.

The opinions farmers held toward genetically engineered seed and other new chemical applications were, on balance, a mixture of general enthusiasm tempered by sober awareness of the costs. While the costs included concerns about potential damage to the environment or food supply, these technical considerations were if anything rivaled by an awareness of how farm life itself was being affected.

Chemicals and GMOs were a bit like the proverbial temptations of the devil that were too alluring to resist and yet a bargain that involved an inexorable price. They made it possible to stay in farming but drove up the input costs to the point that expansion was almost necessary to break even. That did not bode well for the long-term prospects of family farming. In the short run, it was more difficult to know what information could be trusted and to feel that a person was still in control.

"Terrifically seductive" was how a cotton-belt farmer in his late sixties put it when asked about new GMOs and related technologies. Higher yields and less time cultivating the fields were certainly attractive, he thought. But he also felt caught in a kind of cycle that was beyond his control. Fewer people were farming. Fewer knew how to farm. The input costs forced farms to be bigger and bigger. That required more technology. And the cycle continued. "It's like a tsunami," he said. "Whether you're for it or against it, it's there."

SELECTIVE BREEDING

Among the livestock farmers we interviewed, the counterpart to GMOs was selective breeding. An important part of farming for centuries, selective breeding has become more specialized and technologically sophisticated in recent years. Artificial insemination and in vitro fertilization are among the ways in which breeding has become more selective.

Several of the farmers we interviewed described the success they had experienced with these technologies. Semen from purebred livestock provided opportunities to breed animals that produced leaner meat or that matured sooner. In vitro fertilization was a solution for declining pregnancy rates in dairy cattle.

A dairy farmer we interviewed was currently earning as much money selling calves produced through in vitro fertilization as from milk. At the time we talked he had nearly two hundred pregnancies in his herd from in vitro fertilized embryos. The process involved extracting the eggs, taking them to a laboratory in another state, where they were fertilized, and then implanting them in the cows. The cost to create the embryos was $15,000 to $20,000 a month. Each calf produced sold for between $15,000 and $40,000. Genetic testing was being done to identify the DNA most desirable for high-producing dairy herds.

The reluctance farmers expressed about this kind of technological innovation was mostly on economic grounds. As one farmer put it, "I'm a tight-wad." He would invest in something this expensive, he said, only if it was proven to work well. They did not consider it unethical to adopt new technologies simply on grounds that it was somehow wrong to tamper with Mother Nature. They generally trusted the science involved and the testing and regulations associated with new technologies.

The trust farmers we spoke with expressed in these new technologies was in most instances based on firsthand experience as well as information from conversations with specialists and neighbors. A certain amount of self-interest and faith in science and technology was involved. At the same time they respected the need for careful testing and regulations and understood the value of maintaining the public's confidence.

"We protect our animals and our land to the best of our ability," another dairy farmer explained, "because the more we care for the land and the animal, the more production we're going to get out of it." It bothered her when the media portrayed biotechnology as necessarily bad. "Yeah, we need to have [regulations] in place and we need to do the research before anything comes to market. But we're not going to pillage the land or hurt those cows out there. That's just not going to happen."

SUSTAINABLE ENERGY

The area of potential technological development that the farmers we spoke with were least optimistic about was sustainable energy. Because their fuel costs had risen dramatically in recent years, they acknowledged the value of becoming less dependent on petroleum-based fuel. Rising petroleum costs were also a cause of higher input costs for chemicals and fertilizer. They hoped more could be done to promote sustainable energy, but were doubtful about significant progress being made anytime soon.

Several of the farmers we spoke with had tried running biodiesel in their tractors. Their interest stemmed from wanting to help the nation become less dependent on fossil fuel and from hope that the costs would be lower. In the corn belt biodiesel was being promoted as a way to heighten demand for soybeans. But farmers expressed disappointment. "At first everybody said biodiesel, that's soybeans," a farmer who grew soybeans observed. "Well, we find out now it's not really soybeans. It's dead cows and people find a dead deer and they're using that." He was having trouble using biodiesel. "It rusts my fuel tank on the tractors," he said. "You take the fuel cap off and there will be a gooey mess."

Ethanol was the topic toward which most farmers were favorable. This was particularly true among farmers raising corn, but was also the case among some wheat-belt and cotton-belt farmers who had started growing corn. They understood that corn prices were up because of the demand for corn from ethanol plants. They were aware of the criticisms that ethanol was driving up food prices and was perhaps not as efficient to produce as proponents hoped. They nevertheless defended it.

"Everybody's blaming ethanol for high food prices," Neil Jorgensen remarked. "What we need to do is educate the media to understand it." For every bushel of corn that went into ethanol, he said, a third of it actually went into ethanol itself and the other two-thirds was by-product used to feed livestock. Besides that, he thought it was important to understand that corn was a renewable product, contributed to cleaner air, and helped the nation to be energy independent.

Farmers we spoke with who were less sold on ethanol expressed doubt that it was quite as efficient as proponents claimed. They worried about the amount of energy required to produce ethanol. It was a short-term solution that was boosting corn prices, but might be replaced by other renewable energy sources.

Methane was another source of renewable energy that some of the live-stock farmers had investigated. They were sensitive to environmental concerns about the effects of methane as well as complaints from neighbors about the odor from feedlots and manure. Several of the farmers we interviewed had investigated purchasing a methane digester that would reduce odor and supply energy that could be used on the farm or sold to utility companies. To date their investigations had not persuaded them that the return on investment was sufficient.

Wind energy was of interest as well, but farmers' opinions toward it were mixed. Clay Jorgensen was one of the farmers we spoke with who knew a great deal about wind energy from having served on a regional commission for sustainable energy. He strongly favored wind energy and was proud to have promoted some projects in his area well before the concept gained wider popularity. At the same time his experience had left him with concerns. The big difficulty was establishing the necessary grids and convincing the electrical companies to invest. When the electrical companies did get involved, the second problem was that farmers resisted, at least if the towers were being placed on cropland. Farmers complained about having to work around the towers, not being able to plant crops within certain distances of them, and having maintenance workers ripping up their fields. Mr. Jorgensen's current view was that wind towers were not a good deal as far as crop farmers were concerned.

That opinion varied of course from region to region. The part of the cotton belt where Mr. Rayburn farms is flat enough that wind could blow for two hundred miles with hardly any interference from anything. One of the largest wind farms in the state is an hour away. A few years back a company came through purchasing leases for future turbines. Nothing in the immediate area has been built yet. It probably will be another four or five years before the transmission lines are in place. Mr. Rayburn has leased some of his land. He likes the idea of renewable energy and hopes to benefit from the leases once the turbines come on line.

Farmers who took a not-in-my-backyard attitude because of the noise expressed the most negative sentiments toward wind energy. There was a constant whip-whip noise every time the huge blades turned, a corn-belt farmer who lived a quarter mile from one of the turbines said. It went on day and night whenever the wind was blowing. He had not anticipated it being so loud.

The greatest enthusiasm for wind energy did not come from farmers who stood to benefit personally for turbines on their land. It came rather from people convinced that sustainable energy was essential for protecting the environment and for achieving national energy independence long-term. One farmer who firmly believed in wind energy had a suggestion for people who opposed the idea. "It wouldn't be a bad idea for these anti-wind guys," he said, "to have them sitting in the dark for a while freezing their asses off."

AMBIVALENCE

The best single-word summary of how the farmers we spoke with felt about technological change was "ambivalence." Their mixed feelings reflected more than simply the fact that bigger machines were nice but costly or that new chemicals produced higher yields but drove up input costs and might have harmful side effects. The ambivalence stemmed even more deeply from their understanding of what it meant to be a farmer and to have a special relationship to the land.

Technology was having mixed effects on farmers' family relationships. It was making it possible for some farmers to be more efficient and to farm more land. It meant that farmers who could help their offspring get started in farming were doing so by sharing land and machinery. But other farmers knew their sons and daughters would not be able to farm. It would be impossible to accumulate enough capital to purchase expensive machinery or to obtain land. For those farmers, newer and more expensive technology did not imply the demise of family farming in general, but it presaged the end of a valued way of life for their own family.

New technology affected farmers' day-to-day relationships with their spouses, children, and neighbors. On the positive side, it was possible to stay in contact via cell phones and it was easier to stay connected with retired relatives who lived in town or with children who lived in another state. Nuclear families could handle larger acreage by themselves by relying more heavily on more powerful equipment, no-till methods, GMOs, and herbicides. On the negative side, the farmer was more likely to be working fields farther from the farmstead than in the past. It was harder to involve children when machinery was complicated and chemicals were expensive and potentially dangerous. Specialized information came from a professional agronomist rather than from informal exchanges over coffee or at the grain elevator.

Farmers' sense of independence was changing as well. New technology made it possible to be more independent in some respects. A farmer with a huge tractor did not have to rely on hired help. A farmer could take pride in keeping up with new developments in genetic modification. Instead of knowing how to fix a plow, being one's own boss now implied understanding how to negotiate futures markets and having good relations with the local agronomist.

New technology nevertheless reduced farmers' sense of independence in other ways. No longer was it possible to grow one's own seed and simply take it to the co-op to be cleaned and treated. Farmers now had to purchase seed from Monsanto and sign agreements about how it would be used. As one farmer observed, "Farmers have been genetically modifying crops and livestock since the dawn of time. The only problem with these genetically modified grains is that they are all copyrighted and owned by agribusinesses somewhere."

Similar ambivalence pertained to farmers' relationship to the land. Large self-guided tractors made it possible to farm more land. They made it possible to conduct business by telephone or surf the Internet from the tractor instead of focusing on the land. Global positioning provided better information about each parcel of land, and chemicals made the land more productive, even though it worried farmers that the chemicals might be damaging the land.

The ambivalence in farmers' characterizations of technological innovation was especially evident in their impressions of where family farming was headed. They imagined an almost utopian world in which technology would continue to advance and improve. Within the realm of imagination were such possibilities as robotic-driven machinery that would permit farmers to avoid fieldwork altogether and genetically engineered seed that would dramatically increase yields and render irrigation unnecessary.

These were utopian possibilities that farmers whose families had survived previous waves of technological innovation could imagine. But at the same time they acknowledged uncertainties about the future of family farming in their communities. They thought family farming, for the most part, would continue but be available to fewer and fewer families.

New technologies were beneficial enough that farmers made as much use of them as they could. The new opportunities challenged traditional values associated with farming but did not fundamentally threaten those values. Farmers developed new narratives that showed how traditional values could mostly be preserved.

New narratives about the relationship of technology to traditional values were especially interesting. Farmers who embraced technological innovation considered themselves judicious in what they adopted and what they did not adopt. Like their parents and grandparents, they tried to outmaneuver the latest fads rather than simply following them.

The stories we heard about technology portrayed the storytellers as decision-makers who bobbed and weaved to avoid the seeming inevitability of rising prices. No-till farmers prided themselves on having found a way to reduce fuel and machinery maintenance costs. The advantage of weed-resistant seed, farmers said, was that it reduced the cost of herbicides. Some of the farmers were annoyed that fertilizer production seemed to be under a near monopoly. They were fighting back by planting nutrient-enriching crops that reduced the need for fertilizer. Others were spending more time monitoring the soil to learn exactly what it needed. Still others were using information technology to budget more precisely the correct input costs.

Ambivalence is the key to understanding how such narrative constructions work. Being ambivalent is not only a matter of being of a mixed mind. It is not simply the capacity to see two sides of an issue. It is rather the ability to formulate those opinions as coherent discourse. Coherence implies connections with timeworn values and with modified interpretations of those values.

Nowhere was it more evident that the information technology revolution of the past quarter century was bringing new narratives into view than in farmers' interpretations of the basic values of family farming. Farms, farmers said, were still good places to raise children and to be one's own boss. True, farm children might not feed the cows or drive the truck at a young age, but they were living in the country and watching their parents operate a complex business. Technology was enabling farmers to do more things than in the past, but previous generations had also found it necessary to adapt. It was taking more capital and more knowledge to manage the farm. That was challenging, but possible. Farming had never been easy.

Even the physical work remained challenging. Better equipment made it easier to handle these manual tasks, but they were still part of the job. An offhand comment from a corn-belt farmer drove home the point. He was not complaining. Although his farm was modest in size, the planting and harvesting went quicker than ever before. The three-bottom plow he started with had long been replaced. He comes in from a day on the tractor. His jeans and shirt are clean. The grain bins have mechanical augurs.

But the grain bins still have to be cleaned by hand. The last of the grain has to be scooped and the bins swept. Yesterday when he was doing this it was 96 degrees outside. "I don't know how hot it was inside," he noted, "but the sun was shining on the metal. It was pretty hot. At sixty-six I'm not real excited about that anymore."

His words are a vivid reminder that farming remains labor intensive at least some of the time, even though, as many of the farmers we spoke with emphasized, it has also become management intensive. For all the worries they have about the effects of technology on their way of life, the attractions of technology are clearly evident.

MARKETS

7

International markets are certainly making life a lot different. Like your little jar of pickles. Most of those are from India. It's cheaper than to produce them here.

—Truck farmer, female, age 54

Right now, Cargill is buying all the seed they can buy and putting it on a train and shipping it to another state. That's raised my feed price dramatically.

—Cattle farmer, male, age 39

The most challenging aspect of farming, farmers say, is having no control over markets. Markets are even more frustrating than Mother Nature. The weather at least is beyond anyone's control. Markets can be manipulated—and are manipulated, farmers think—by human forces that do not have farmers' interests at heart.

This has probably been true for as long as anyone can remember, but many of the farmers we spoke with thought the problem was getting worse. Markets that had always been influenced by international forces were now even more directly shaped by those forces. Prices that used to fluctuate mildly with changes in supply and demand now seemed to fluctuate wildly. And prices seemed to be more influenced than in the past by large corporations.

Ironically, the complaints we heard about markets were ones farmers were articulating at a time when the prices of most farm outputs were at record highs. In some instances farmers were selling commodities at prices three to four times higher than only a few years earlier. They acknowledged that things were pretty good. Over the long haul they figured global demand for farm products would steadily increase. Population was growing, trade was increasing, and incomes even in much of the developing world were climbing enough to give people more capacity to purchase food. The challenge long term was to meet the demand. As one farmer put it, "We have to increase our production, no ifs or buts about it. That is the bottom line."

If long-term prospects were good, the realities facing farm families from day to day and from year to year nevertheless remained mentally and emotionally challenging. For the farmers we interviewed the challenges required making hard-headed decisions about what crops to plant, which technologies to invest in, when to buy and sell, whether to focus on particular niches in the market, and how to think about government programs and regulations.

These were decisions that could be thought of purely in financial terms. But there was a human side to them as well. They affected how farmers thought of themselves. Were they really doing the best they could in caring for their families? Was farming in their blood to the extent that they were staying with it at a high cost to themselves and their families? Was the notion of being their own boss a fiction? Could they take pride in being able to adapt to changing conditions?

These were the questions we wanted to address in listening to farmers talking about their decisions in negotiating markets. The farmers we spoke with paid a lot of attention to shifting market conditions. Many of them received market reports four or five times a day. They went to meetings sponsored by farmers' organizations and talked knowledgeably about the decisions they had made that had or had not worked to their benefit.

It was in this context that some of their most revealing comments about the meaning of farm life occurred. They felt that markets were inherently beyond their control and perhaps were even stacked against them. And yet they also described the sense of accomplishment they derived from finding small ways in which to fight back and in these ways to regain some control.

AT THEIR MERCY

The way most of the farmers we spoke with talked about markets was similar to how they talked about Mother Nature. Neither could be controlled. The difference was only that nobody expected to be able to control Mother Nature. Markets were human and thus were conceivably controlled by someone. Or influenced. Or subject to influence. It was just that those influences rarely involved anything farmers did or could do.

Third- and fourth-generation farmers knew there was nothing new about markets being beyond their control. If they had a sense of farming history, as most did, they understood that farmers had always been "the little guy." The image was of individual farmers tilling small plots of America's land, hauling their crops and produce to town, and having no say at all in how much they got. These were the wellsprings of populists' and agrarian progressives' complaints about railroad prices and banking monopolies.[1]

It went without saying that impersonal laws of supply and demand determined how much they got. The big grain elevator chains, the railroads, the Chicago Board of Trade, and lawmakers in the nation's capital shaped it as well. They knew this from family legends and from firsthand experience. And yet, their inability to set prices or to do much that would affect these markets left them feeling weak, vulnerable, and frustrated.[2]

Across the corn belt, markets had mostly been fluctuating upward for the farmers we interviewed. "Soybeans were good this morning," a farmer observed, noting that the price was a third higher than the previous fall. Good markets could easily turn bad, though. It was hardly being pessimistic to say so. Farmers knew this from personal experience.

"You can't judge what's going to happen in the markets," the same corn-belt farmer observed. "There's going to be times you make good choices. There's going to be deals you make that are losers." He recalled an old farmer telling him that years ago.

The reason for farmers' worries even when prices were high was not only that prices could fall. It was also the perception that markets were still rigged against them. No matter how good the prices were for grain or livestock, the input costs seemed to rise just as much or more. Net income seemed to remain small.

A wheat-belt farmer who farms 7,000 acres of mostly rented land and earns most of his income from wheat and cattle provided an interesting illustration of how even some of the more successful farmers we interviewed thought about markets. Although the scale of his operation is one of the largest in the area, he says the margin on which he operates is slim.

"As grain prices rise," he says, "the input prices rise right along with them." And when grain prices taper off, "the other prices don't." Recently the cost of fertilizer and equipment rose at the same rate as grain prices. But then the cost stayed high when grain prices fell. "The wheat price right now is two dollars a bushel lower than it was shortly after harvest," he observed, "but none of those other prices have dropped back. And they're not likely to."

The cotton growers we interviewed were generally dissatisfied with the market conditions in which they purchased supplies and in which their cotton was sold as well. Those who had been in the business for three or four decades remembered when most of the cotton they produced was used domestically. Now they figured nearly all of it was sold in international markets. They thought the United States could do better in shaping those markets.

"The only country I know where a foreign interest can come in and influence public policy is the United States." That was how Mr. Rayburn saw things. "We give in to Third World countries when they come crying on our shoulder." He conceded that some efforts in what he regarded as the right direction were being made. But he thought a lot more needed to be done.

Most of the dairy farmers were dissatisfied too. Their problem was not global competition. Milk was sold in domestic markets. The difficulty was that input costs were rising but milk prices were not. Or at least not by as much.

"As soon as milk starts going up a little," Mr. Loescher complained, "you always have a middle man whose hand is out first. His pockets get bigger while ours usually stay the same." Larger dairy farmers shared this view. Although they were better positioned to deal with middlemen, they were experiencing lean years as well.

Farmers are stereotypically dour. Comments like these suggest that the stereotypes are apt. They fit the stereotypic farmer who greets a beautiful morning with "We'll have to pay for this!" But feeling that a person is at the mercy of fluctuating markets is not the same as taking a pessimistic view of everything. Perhaps because our interviews were conducted anonymously, many of the farmers we spoke with acknowledged that farming was good in many respects. The work was easier. Yields were up. Still, they knew that they had little control when it came to markets. They were small players in that arena.

Being the little guy in big markets, farmers recognized, was similar to the plight of individual workers competing for wages in big manufacturing sectors such as the automobile industry. For industrial workers the answer had been collective bargaining through trade unions. Some of the farmers we interviewed belonged to farmers' unions. Their experience was generally less than satisfactory. The unions functioned like professional associations. They did not engage in collective bargaining. They provided information and organized meetings. They did some lobbying and helped with public relations. They were similar to other organizations to which farmers belonged and supported through small contributions.

The principal difficulty these farm organizations faced, according to the farmers we spoke with, was that farming's diversity led more often to disagreements than to agreements. Mr. Rayburn had held offices in several statewide and national cotton growers' associations. He thought these organizations were definitely beneficial and that farmers should be actively involved in them. As markets became more internationally competitive, he considered it increasingly important to support organizations that advertised and promoted farm products in those markets. But anything dealing with more specific farm policies, he thought, was unlikely to be effective.

The examples Mr. Rayburn had in mind were regional differences among cotton growers. Some of the growers farmed close to major ports, while others depended on long-distance rail shipping. Some farmed only a few hundred acres, while others farmed several thousand acres. They had different views about technology and genetic modification and chemicals. "The diversity of the issues that we have to deal with," he said, made it difficult to find common ground.[3]

Regional diversity meant that smaller associations typically worked better as sources of specific farm management advice. These included co-ops, local growers' associations, and extension services. One of the wheat-belt farmers we spoke with described the farm management association he looked to for such advice. The association linked farmers with agricultural economists at one of the state universities. He turned to them when making decisions about equipment purchases, land, taxes, and farm loans.

As an individual farmer, the difficulty he most often faced was lacking information. By pooling information from the hundreds of farmers it dealt with, the management association provided that information. "If I want to trade a piece of equipment," he explained, "I will call them and say this is what I have and this is what the dealer wants to trade for." He asks if they think he is getting a fair deal. They look at all the information they have and tell him how the deal he's getting compares.

MARKET FLUCTUATIONS

Many of the farmers we spoke with complained that market prices for grain and livestock fluctuated more from day to day than in previous decades. "When I started farming maybe for the whole month they would move only a penny or two," a farmer in his late sixties remarked, "but now, like today, corn and beans are off twenty cents over yesterday."

A wheat-belt farmer expressed a similar view. "Nowadays, it's easier to make a poor decision than it used to be because the economy is just crazy," he explained. "We're world-oriented now," which meant that markets were affected by events he had no way of hearing about. "But all the big guys in Chicago hear about it and they're going to change the way things are done like overnight." He felt it was hard to avoid getting "caught in the squeeze" when markets were so unpredictable.

That was true among the truck farmers we interviewed as well. They were less likely to experience dramatic price fluctuations from day to day. But they considered markets to be increasingly unstable. The price of carrots or apples, they thought, was harder to predict than in the past. Prices seemed to jump or fall in ways that had little to do with local production.

Farmers were less clear about the reasons for these fluctuations. The most common reason, they thought, was the fact that agriculture was more influenced than ever before by world markets. Some of the farmers we spoke with also thought traders were victimizing them. "These people who are professional marketers," one farmer complained. "They don't understand it either." And in his view the professional traders were causing fluctuations because of speculating on the price of commodities.

The way to cope with market fluctuations was to hedge by trading in commodities futures. Most of the larger farmers we interviewed said they did trade in futures markets. Many of them worked with financial consultants who they hoped understood futures markets better than they did. A few of them acknowledged following the futures market almost as an obsession. "Oh, twenty or thirty times a day," one farmer chuckled when asked how often he checks the markets.

The prevailing view was that hedging did help to stabilize income in fluctuating markets. The downside was figuring out how much to hedge at and what price. "We've been turned into gamblers," was how one farmer put it. No matter how one played the market, there was a good chance of losing money.

But even the farmers we spoke with who were doing well financially and who spent the most time following the markets complained that this was still the aspect of farming they disliked the most. "I hate grain marketing," one farmer exclaimed, "because I do not believe it's anything but a random number." He said he had participated in a study in which various market specialists gave advice over a five-year period and none of them got it right more than two of the five years. "It just bugs the heck out of me," he added. "With price swings, it has a huge impact on the bottom line, but it's a random number."

Mr. Bower was one of the farmers who talked at some length about his understanding of market fluctuations. The two factors completely out of his control, he said, were the weather and international markets. He understood that the international market—for wheat in his case—was driven only partly by supply and demand, which in turn was affected by the vagaries of weather, such as a drought in China or a bumper-crop season in Russia. The markets were also shaped by the human decisions made by governments. And that was especially frustrating because governments' ways of intervening in the market were not only complex but also capable of producing unintended consequences. Indeed, government policies affecting agricultural markets were so complicated that it was difficult for farmers to understand them, let alone do anything to influence them.

The Armstrongs were another family that felt it was increasingly difficult to deal with market fluctuations. Three of the last five years had been good financially, but the last two had barely covered expenses. Operating close to the break-even margin elevated the importance of market fluctuations as small as 2 or 3 percent. But wheat and soybean prices had been fluctuating by 30 to 40 percent. Last year the Armstrongs hired a consulting firm to help watch the market and give advice. But that still did not put things within their comfort level. "I don't like the feeling of always having to be watching it," Mrs. Armstrong asserted. "You go plant or harvest and it's done. But with marketing, it's always hanging over your head."

MARKET STRATEGIES

The traditional way in which farmers adapted to changing market conditions was to shift from one kind of output to another. For example, many of the farmers we spoke with had grown up on farms in which eggs and poultry were a regular source of income, but few were still raising and feeding chickens. The reason was that large agribusiness corporations more efficiently produced eggs and poultry than average-sized farms. The same was true of hogs, which a number of farmers we spoke with had raised until the 1980s, when hog production ceased being profitable. Farmers specializing in grain production had shifted within the limits of soil types and rainfall from varying amounts of wheat, milo, corn, or soybeans.[4]

Shifts of this kind were driven by economic considerations, but they were legitimated by family legends. Farmers who grew wheat on the same land where their parents and grandparents grew wheat took pride in carrying on a family tradition. Farmers who now raised something else took similar pride in being flexible the same way that previous generations had been.

As one farmer explained, he switched a few years ago from growing sugar beets to raising corn and soybeans. The shift was similar to his father getting out of the cattle business and raising sugar beets. Being willing and able to adapt was, in this farmer's view, a mark of good farming. It was part of his family's tradition. One of his neighbors grew potatoes until mechanical harvesting made it impossible to do that efficiently in heavy soil. He now plants mostly corn and soybeans. He considers it part of dealing with Mother Nature to make such changes.

Another farmer shared a similar understanding of the need to be flexible. His maternal grandparents were in the chicken business. One evening in the 1940s they heard on the evening news that someone had invented refrigerated trucking. The grandfather turned to the grandmother and said, "We're in trouble." Within a few years they were no longer able to compete. The Tyson Company in Arkansas was raising chickens by the tens of thousands and shipping them to urban markets at unbeatable prices.

How farmers framed their accounts of the changes they had made was as important as the changes themselves. The question in farm management studies has always been why some farmers adopt new methods and others do not. Part of the answer is capacity. Farm size, for example, matters. But the missing piece is that farmers also have to feel a sense of efficacy. The family legends remind them that change was possible in the past. The accounts

169

provide ways of emphasizing that farmers are not entirely at the mercy of market forces. True, it may have been an economic necessity to switch to different crops. But it matters to be able to emphasize that choice was still involved.

The point of the Tyson story was not that the grandparents had been victims of changing market conditions. It was rather that the grandparents realized early that times were changing and switched to a different kind of farming. The storyteller drew a lesson from their experience. The lesson was to avoid letting your self-worth get too attached to the soil or to one kind of farming. A person needed to have enough self-confidence, the farmer said, to go in a new direction when times changed.

Choices made even under difficult conditions reinforced the sense among farmers of being one's own boss. A good boss knew from family traditions that change was inevitable. A person could take pride in having made the change at the right time or in having been the first farmer in the area to try something new. A good decision-maker could also take pride in having learned how to benefit from a new technology. The transition went more smoothly when it could be interpreted in these ways.

The family stories that legitimate switching to new crops contradict the stereotype of farmers being too wedded to the past to change. To be sure farmers do many things the same way year after year. And yet they have narratives in their repertoire that say, in effect, our ancestors adapted to changing conditions and we can too.

The changes farmers interpret as choices are nevertheless sources of frustration. Although their descriptions of these changes show that being able and willing to adapt is good, the narratives also demonstrate how much farmers appreciate being able to plan. "I like to have all my little ducks in a row," a farmer in his sixties says. But something happens that "changes my whole flight pattern." He thinks maybe he is just getting old and tired, but that kind of thing "bugs me," he says.

The changes farmers described seldom reflected a shift away from greater diversity in the past toward narrower specialization. Those changes had already taken place in previous generations. They were associated with the early twentieth-century shift from family-subsistence farming involving multiple crops and kinds of livestock to single-crop, market-oriented agriculture. The recent shifts were from one kind of output that was already market oriented toward another that was now more profitable because of new technology or different market conditions.

The largest shift in recent years among the farmers we spoke with was in response to the rising demand for corn for ethanol production. In several of the communities we studied farmers who had previously grown only wheat, cotton, or rice were now planting corn. Corn was more profitable because of higher prices driven by the demand for ethanol. It was also more profitable, farmers explained, because of exports to developing countries where rising family incomes were expanding the demand for beef and pork, which in turn increased the demand for corn.

Genetic modification was expanding some of these options. For example, one of the wheat-belt farmers we spoke with was still planting approximately two-thirds of his cropland to wheat, but was planting the other third to corn. "Dry-land corn in this area would have been crazy a few years ago," he said, but new seed varieties were available that permitted corn to grow with less water.

In smaller ways many of the farmers we spoke with were experimenting with at least some crops or other activities that gave them greater control over prices. One example was a couple who had shifted from dairy farming to growing pumpkins. It seemed crazy, they said, but even though people really did not need pumpkins, they were willing to pay high prices for them for Halloween decorations. Another example involved a shift from truck farming to selling hay. The farm was located in an area where more people were boarding horses and thus in need of hay.

These strategies were risky, though. One of the farmers who switched to hay observed that he still had to compete with hay shipped in from Canada where, he thought, the Canadian government was unfairly subsidizing farmers. Other farmers nevertheless felt it made sense to produce output that could be sold in local markets, such as pumpkins and hay, rather than being at the mercy of international markets.

The larger point farmers emphasized was that market strategies are less about specific activities and more about a person's general outlook. Although there were ways to strengthen one's position and to gain some control, markets were always going to be to a considerable extent unpredictable and beyond any one person's control. That meant having a view about how much risk to take.

A common perception of farmers is that they are generally risk averse. We spoke with a number of farmers who fit this stereotype. They described themselves as fiscal conservatives who never went into debt and never did anything that might backfire financially. These farmers often had hard-luck stories about parents, grandparents, other relatives, or neighbors who had suffered the consequences of risky decisions. In their case the family legends were cautionary tales.

It was interesting for this reason to find an even more common view that favored taking risks. This view suggested that farmers unwilling to take risks would be left behind. Risk was necessary to expand one's holdings, to purchase new machinery, and to experiment with new technology. Farmers holding this view said it was simply good business to take out loans and to go out on a limb from time to time. To succeed most of the time, they considered it necessary to risk failing some of the time.

How they thought about risk reflected family traditions. Farmers who favored caution had stories to tell of ancestors and neighbors who had learned the hard way to be cautious. Risk takers pointed to good decisions their parents or grandparents had made which demonstrated the value of taking risks. The stories legitimated a general and apparently enduring orientation toward risk.

But there were also indications that farmers were thinking about risk in new ways. The added information provided by GPS and from soil science, coupled with genetically engineered seed varieties, was giving farmers opportunities to make more decisions about specific ways of handling risk than in the past. It was possible not only to hedge markets by trading commodities futures but also to bet in small ways against the weather.

Mr. Hebner provided an interesting example. Although wheat is his primary crop, he also plants several hundred acres of milo. Each year he has to decide which seed variety to plant. Last year he selected what he calls a "full-season milo" that does really well in a good year when weather conditions are just right. "I liken it to swinging for the fence when you're playing baseball," he explained. "If you hit the ball, you'll probably hit a home run. You'll either hit a home run or strike out. You probably won't make a double." The seed variety he selected was wrong. The weather did not cooperate. "I swung for the fence and I struck out."

What did he learn? The lesson was not to avoid risks at all costs. He hates taking risks. They are the part of farming he especially dislikes. At the same time being able to take risks and live with the consequences reinforces his sense of independence. Being up at bat is a fitting metaphor. It is your show at that point. You alone decide whether to swing hard, bunt, or hold. The lesson from taking risks is that striking out some of the time is part of playing the game.

NICHE MARKETING

The additional strategy that many of the farmers we spoke with used to strengthen their position was niche marketing. They realized that producing interchangeable commodities no different from every other farmer in their area left them entirely at the mercy of factors beyond their control, such as shifts in international demand and supply. Doing something different was risky, but could be a way of protecting themselves against these exogenous factors. This was one of the ways in which their penchant for independence came into play. They took pride in bucking the trend, going against the grain, trying something different.

The purebred livestock farmer we heard from in previous chapters who owns ten thousand acres and sends exports to Latin America is an example of a niche farmer who took a risk that proved highly successful. As a young farmer he tried without much success to make a living milking cows and raising hogs. The idea of raising purebred livestock was virtually unknown in his community. He gambled and spent $2,500 on a purebred Italian-born bull. Over the next five years the bull produced a million dollars' worth of semen.

A farmer in another community who raises corn, soybeans, and wheat but earns most of his income from running about five hundred head of livestock on grassland described a different niche market. He specializes in breeding high-quality Gelbvieh bulls. He prides himself in following his father's innovativeness in experimenting with the golden brown breed that started to be imported from Germany in the early 1970s. He thinks other farmers have been too easily swayed by the popularity of Angus livestock. He figures it makes more sense to be doing something different.

A family near the Loeschers was supplementing their dairy income by breeding better Holsteins. Times were difficult when we talked with them. Milk prices were 20 percent lower than twelve months ago. Rainfall was well below average. The previous year they lost a third of their herd and all their corn when the area flooded. "I kind of wonder why we're in this business. We need to have our head examined," the wife lamented. But they had no intention of quitting. A few years ago they tried selling seed but were undercut by one of the big agribusiness companies. Breeding calves for sale is working better. Genetic testing and embryo transfer make the process more effective. About 15 percent of their income now comes from merchandising the animals.

On a smaller scale several of the farmers we spoke with had begun raising goats for meat. The idea had been practiced mostly by hobby farmers, but with greater interest from consumers and more opportunities to engage in direct distribution, farmers with modest incomes from crops and with time available were finding this an effective way to avoid going under.

Another example of niche marketing involving livestock illustrates the possibilities for farmers to engage in vertical integration. One of the farmers we interviewed was part of a producers' cooperative. At first the producers raised calves and sold them to a feedlot that fattened them and sold them to a meat processing plant. The producers then vertically integrated. They pooled their resources, purchased a large feedlot, and founded a producer-owned meat processing plant that sells in international as well as in domestic markets. The return proved to be very substantial, the farmer said.

These were not the only examples of farmers profiting from becoming more directly involved in international markets than anyone in their families previously had been. Another example was a wheat farmer who was an officer of a wheat growers' association that played a role in developing certified seed engineered for particularly arid land in the Middle East. When the ruler of one of the Middle East countries in which the seed was being used decided to privatize some of the royal land, the farmer and several other members of the association bid successfully to secure the lease.

The only condition was that they introduce high-producing, certified seed and the latest developments in fertilizer. Language was initially a barrier, but the farmer found a translator living in another state who was also doing business in the Middle East, and together they make periodic trips to oversee the operation there. This farmer still grows wheat on some of the land his grandfather farmed. He still does business at the local co-op and chats with the neighbors over coffee. But he is part of a growing trend among American farmers to be directly involved in overseas operations.

The niche market that has become increasingly important over the past half century is organic farming. Every farmer we interviewed had an opinion about organic farming. But only a few had tried it. Most viewed it as a niche that would not be reasonable to pursue in their own case. Many were skeptical about the concept of organic farming. They saw little value for consumers from purchasing organic goods. Others said they were interested, but the investment to get started was prohibitive. Besides the time involved to get the land properly certified, on-farm storage was often required because the larger co-ops and elevators were inadequately equipped.[5]

Some of the dairy farmers in the Loeschers' community were producing organic milk. The Loeschers had thought about it, but they were operating on too small a margin to make it work. Perhaps it was sour grapes, but they were also skeptical that organic milk was as pure as the label implied. Some of the crops being used to feed cows organically were being fertilized with manure from cows that were not being fed organically. Another source of contamination was milk truck drivers who, the Loeschers suspected, sometimes topped off a tank of organic milk with regular milk.

Several of the Loeschers' neighbors were successfully producing organic vegetables, mutton, and wool through a community-sponsored agriculture initiative. The initiative was functioning on a fairly small scale but was working because townspeople were interested enough to become involved. The program was a cooperative plan in which consumers joined the co-op as members and promised to purchase a specified amount of the farms' output each year. A side benefit was that some of the farms were able to expand by installing solar-heated greenhouses.

"It's a new niche," one of the community leaders involved in the effort observed. She thought it was likely to expand. The nonfarming public was interested in eating healthy, and enough of the local truck farmers were benefiting that they were able to make it work.

Organic cotton was being grown by a few of the farmers in Mr. Rayburn's area. He called it a niche market, by which he meant there was a small segment of the consuming public that wanted it. "But let me tell you my feelings on organic cotton," he said. "You produce less. It's more labor intensive." That made it harder for farmers to make money on it. Consumers also had to be willing to pay more for it. His philosophy was to invest less labor and capital in what he produced. Not more.

Although several of Mr. Rayburn's neighbors were growing organic cotton and apparently making money doing it, he doubted that the cotton could still be considered organic by the time consumers purchased garments or towels made of it. "I have yet to see a cotton fabric that's organic that's not been dyed with some kind of dye and therefore is not organic anymore."[6]

Organic farming was but one example of the niche markets farmers we spoke with talked about. Other examples ranged from inventing a new piece of equipment in their farm shop that they could sell to other farmers, to experimenting with new seed varieties or kinds of fertilizer, to developing a specialized kind of business or technology.

A cotton-belt farmer we interviewed provided an illustration showing how technology and social networks can lead to a profitable niche market. She and her husband had a friend in college whose father purchased a delinting machine that processed cottonseed to turn it into planting seed. The father's plan was to set the son up with a small delinting business that would bring the son back to the area. Through the son, the woman telling the story and her husband became partners. The couple farmed as well, but more of their income came from the delinting business.

The delinting story illustrated that niche marketing can be profitable but also entails risks. Within a few years the couple bought out their partner, expanded the business, and started a second one in another location and then a third. But then genetically modified seed came along and the couple's delinting business dropped by 70 percent in one year. They knew it was time to shut down. Over the next five years they shifted almost entirely to raising cattle.

A farmer who had been in the asparagus business provided another example of the fragility of niche markets. Rising demand and cheap labor to do hand harvesting had encouraged him and his neighbors to shift land from other truck crops into growing asparagus. The business was highly profitable for several decades. Then the US government entered into an agreement with Peru and Chile aimed at discouraging the production of cocaine in those countries. The agreement subsidized asparagus farming in those countries. It essentially shut down the US asparagus industry.

The other niche markets farmers described were equally precarious but in the short-term provided sufficient revenue to keep farmers from going under. Selling hay to local boarding stables became difficult when economic conditions fell so sharply that people with horses could no longer afford them. Another farmer bought and sold machinery. The niche was there

175

temporarily because the local implement dealer had gone out of business, forcing farmers to drive thirty or forty miles to a larger town. The niche was available, though, only as long as the farmer was able to persuade the wholesaler to supply him with machinery.

Other strategies for supplementing farm income were closer to traditional part-time off-farm jobs. A wheat-belt farmer supplemented his regular crop income by assembling machinery during the winter in his farm shop. A small dairy farmer kept a trucking business going on the side, hauling hay and grain for other small farmers in the area. A truck farmer operated a farm market that sold both homegrown and imported items. These were activities that generally fit well with on-farm seasonal work. In statistical reports they still counted as on-farm work even though they were similar to the off-farm work a farmer might have done in town. They did not provide a stable income with health insurance and retirement benefits, though. That was the reason many of the farmwomen worked at off-farm jobs.[7]

EFFICIENCIES OF SCALE

None of the farmers we spoke with thought it was ever possible for farmers individually or collectively to become large enough to shape the markets in which they sold their outputs. This view was part of their reasoning about being at the mercy of markets. They did insist, however, that larger units of production had an advantage in competitive markets. They felt that larger farms were inherently more efficient than smaller ones—meaning lower input costs relative to output and thus higher profit margins.

The dairy farmers we spoke with were among the ones who were most convinced about the efficiencies of scale. The ones with smaller herds, like the Loeschers, knew they were at a competitive disadvantage with large dairies. The larger dairies in their community were doing better in dealing with tight markets and low prices for milk.

John Stratton is a farmer in his late sixties who started with just a few cows and a small plot of inherited land. Today he and his wife operate one of the largest dairy farms in their area. They currently have about a thousand milk-producing cows and nearly that many heifers or younger stock. The farm sells about 75,000 pounds of milk every day. Last year it totaled more than twenty-five million pounds. Counting the land necessary to feed the herd, the equipment, and the value of the cows, the capital investment involved exceeds ten million dollars. Mr. Stratton spends most of his time handling the financial aspects of the business. It takes two hours most days just to pay the bills, another two hours to monitor the herd's performance, and several more to discuss business with neighboring farmers who supply grain on contract and the financial agents who broker loans. Two dozen employees do the milking, feeding, and other physical tasks involved.

The large scale of Mr. Stratton's farm enables him to achieve several kinds of efficiency. One is cost-per-cow for forage. The average in his area is two acres per cow. He has reduced that average by about 15 percent by purchasing additional digestible fiber. A second source of efficiency involves spreading the cost of milking equipment over a larger herd. He also contracts with other farmers in the area for grain and is able to reduce or at least stabilize costs by investing in the futures markets. With all of that he has still been experiencing a negative cash flow in recent months, but he expects to hold on until market conditions improve.

In support of the argument that larger farms are more efficient, other farmers pointed to economic reports they had read or heard about but mostly cited anecdotal ideas, such as the conviction that with the same amount of personal labor they could farm more land than they currently did. The arguments were not always airtight—for example, some of the farmers who believed in efficiencies of scale also thought the ratio of net income to gross income was constant (at about 15 percent) for small and large farms alike and others suggested that per-unit management costs actually increased as farms got larger.

One farmer who had paid particular attention to the economic statistics concluded that the argument about efficiencies of scale was actually false. It was false, he said, because some small and medium-sized farms were efficient and others were not. The only difference was that a higher proportion of large farms were efficient. Nevertheless, the generally shared conviction was that market forces were pushing farms to become larger and that farmers' own survival depended on following that pattern.

That conviction posed the related question, though, of large corporate entities, such as corporations with large landholdings or businesses that play an increasing role in agriculture in some other way, such as Cargill and Archer Daniels Midland, ConAgra, Tyson Foods, or Monsanto. The farmers we spoke with were ambivalent about the role of these entities.

On the one hand, true to their view that market forces favored larger entities, they thought these corporations were inevitable and were destined to play an even more important role in American agriculture in the future. Some farmers even saw personal benefits from the activities of these large corporations. "They are not inherently evil because they are large," one farmer observed. Another farmer, speaking specifically of Archer Daniels Midland, said he was a big fan because the company seemed to have the farmers' best interests at heart. He cited an example in which the company had even boosted local farmers' profits by what he considered an extra fifty cents to dollar a bushel.

On the other hand, farmers expressed worries about the growing impact of corporate agriculture. They felt that corporations were compromising the traditional values inherent in family farming. They worried that cash contracts were undermining smaller farmers and that big companies were less interested in preserving family farms than making profits. Large co-ops that had gone out of business provided a cautionary tale. Other businesses could also become large and then fail at farmers' expense.

An area of particular worry was the potential effect of large corporations on the land itself—its prices, its subjective meanings, and how farmers related to it. The farmer who thought large corporations were not inherently evil, for example, also remarked that corporations were "out for their own best interest" and did not care how long someone's ancestors had farmed a particular piece of ground. That difference implied a shift in land ownership and pricing practices away from sentiment and toward a more purely market-driven pattern. Whereas a family farmer would be inclined to hold on to a farm during a period when profits were low or nonexistent, a corporation would be more likely to buy land below market value when family farmers were forced to sell.

Concerns about the effects of corporate farming illuminated an important aspect of farm families' understanding of their relationship to the land. That relationship was special, intimate, like a marital bond, like parents' love for their children. A special relationship like that would be violated, sullied by entities that treated the land only in terms of self-interest. Monetary considerations would take over. The intimate relationship with the land would be broken and other considerations potentially less beneficial to the land would dominate. As one farmer put it, "If it's big corporate farming, I'm sure they won't take care of the land as well. A smaller farmer would tend to take care of it a lot better. Those big ones, all they care is if they can make some money off it."

Other farmers searched for analogies to show the ill effects that would happen if large corporations replaced family farms. Clay Jorgensen thought it would be like the Mafia taking over. Corruption would be rampant like it was in Chicago during Prohibition. He figured Mafia-style farming would also dramatically increase food prices. "People keep yelling about getting the small guy out of farming," he said. But if that happened, it would be like a "Mafia-controlled thing" and people would be paying a lot more for their food.

It was not just that corporate farming posed an economic threat. It was also that corporations defined a different kind of social relationships. However large family farms might be, they still seemed small by comparison to corporations. Corporate life was urban, distant, and impersonal. Farm life was rural, close, and personal.

"Community is important to us," a farmer who did not like the idea of corporate farming explained. "We live here, we own this community, and we're community minded. In the end, that's probably what we all want out of life is a great community to live in." He thought corporate agriculture would destroy that sense of community. "Yeah, they may have a community outreach program, but it ain't nothing compared to what the family farmer can do."

Although they worried about the effects of truly large-scale corporate agriculture, most of the farmers we spoke with were relatively sanguine about the idea of incorporation itself. The reason was that they had become used to family farms becoming incorporated. As one farmer observed, "Most of the corporate farms are families." In his area this was the case because land was owned by multiple generations and those families had set up a corporate structure for tax and legal purposes. He thought farmers who said they did not like farm corporations were probably reacting more to the difficulties of working with multiple family units than toward the idea of incorporation itself.[8]

A farmer in another part of the country shared the view that corporate farms were generally still family farms. The farmers in his area who were incorporated had done so, he felt, to secure their assets against liabilities. Others mentioned tax and inheritance advantages to incorporation. At the same time farmers recognized that small family corporations could lead to larger corporations taking over more of the land in their communities.

Some of the farmers we spoke with also made the important point that larger-scale farming is driven by factors other than considerations about efficiency. Extended families' desire to keep land in the family is one consideration. Many of the wheat, corn, and soybean farmers we interviewed were farming land owned by aunts and uncles and cousins. They were happy to have this much acreage. Another consideration was the advantage of large-scale operations for achieving greater diversification.

A truck farmer whose operation was among the largest of anyone we interviewed provided an illustration of the advantages of diversification. Having land in several different locations offered protection against hailstorms and other local variations in weather. The farm's size permitted it to have specialized weighing and sizing equipment as well as cold storage facilities. It was profitable for smaller farmers in the area to contract with the farmer and to use these facilities rather than purchase their own. The farm's size also put it in a good position to serve a brokerage function between local farmers and some of the nation's largest food processing companies.

FARM SUBSIDIES

In our interviews we saved the topic of farm subsidies until toward the end. Although they had usually been talking for a long time by that point, the farmers we spoke with were eager to express their views about farm subsidies. They knew this was a topic of concern to the general public. It was a topic they considered routinely in thinking about their own expenses and revenue streams.

Not surprisingly, farmers' held varying views of government programs. Views varied depending on where farmers lived, what kind of farming they did, and whether prices had been good or bad in recent years. Farmers understood these variations. In fact, one of their frequent complaints was that government programs too often seemed to follow a one-size-fits-all logic. Farmers felt that government officials were too far away to understand local conditions.

This complaint was directed toward the media as well, especially in its depictions of farm policies. A typical comment was that agronomists or agricultural economists would do a credible study, but then the media would pick it up and suggest that farmers everywhere should be subject to some new government program based on that study. For example, a study might show that a four-year rotation of crops and hay was desirable in a particular area, but some journalist who had little knowledge of farming would propose that all farmers should follow suit. The problem at least illustrated some of farmers' frustration with government policies.

The most consistent view was that farm subsidies as understood during the twentieth century were now less useful than government-supported crop insurance programs. The earlier programs were designed to stabilize prices through guaranteed price support plans and by controlling supply through regulations on acreage planted. The newer programs that many of the farmers we spoke with preferred provided insurance against crop failures. The prevailing view was not that older versions should be abandoned entirely but that some aspects of both should be continued.

The argument for some form of government support that had been in place since the 1930s was still in the forefront of farmers' minds. The argument stressed the nation's need for food and fiber security. It might be cheaper in the short run for consumers to purchase cotton from India or wheat from Russia, but that was putting the nation at risk. Especially if war broke out or trade routes were disrupted or crops failed, the supply of essential food and fiber would be disrupted. Nothing was as detrimental to stable government, farmers who knew history said, than riots driven by food shortages.

Having food and fiber security was Mr. Rayburn's way of explaining why farmers need a government safety net. He knew critics disagreed. "Well, my business doesn't have a safety net," they would say. His reply was that the safety net was less to protect farmers than to protect the nation. Food shortages of the kind Europe faced during and after the two world wars needed to be avoided. Subsidies should be used judiciously, he felt. They should be used when prices naturally fell so low that the nation's farmers could no longer stay in business. They should also be used to protect farmers against unfair competition from other countries.

Another argument for farm subsidies reflected farmers' desire to be their own boss. Perhaps ironically, the argument for depending to some extent on government for help was connected to farmers' valuing independence. The logic was that without government intervention, the natural competitive dynamics of the marketplace would lead to family farmers going out of business.

"If you want the government out of it, then you get the Rockefellers in there," a wheat farmer who favored current farm programs observed. "If you don't have farm subsidies, your small farmers won't be in business and it'll be one big guy who will dictate things." He could imagine big companies taking over and becoming too large for anyone to control except government.

Many of the farmers we spoke with understood that the public viewed the cost of farm programs differently. Because food and fiber were rarely in short supply, the public more commonly thought about farm programs in terms of their cost to taxpayers. Farmers were eager to show that most of these expenditures did not result in direct income for farmers. Instead, the largest share went for food stamps and school lunch programs. It was politically necessary to include these items, farmers said, in order to get farm bills passed. But they wished the public understood this.[9]

Perhaps for this reason the farmers we spoke with frequently voiced complaints about food stamp programs. As one farmer complained, "there's lots of fraud" in those programs. Others' views reflected farmers' emphasis on hard work and personal responsibility. Food stamp recipients were the kinds of people, in this view, who did not work hard and did not take personal responsibility for themselves.

Subsidized crop insurance programs were easier for most of the farmers we talked with to defend than more traditional price support programs. Because crop yields varied so dramatically from year to year, it seemed wise to even out these fluctuations with insurance. It was less clear why government support should help subsidize crop insurance. The answer farmers gave was usually that the cost of insurance might otherwise be prohibitive.

The proposal widely entertained in the nonfarming media that government subsidies of any kind be limited to small farmers drew mixed responses. Nearly all of our interviewees agreed that wealthy movie stars who owned land or Wall Street billionaires who did should not receive government subsidies. That said, nearly all agreed that it was impossible to determine among ordinary farmers where the line between "large" and "small" should be drawn.

There were at least three difficulties, farmers said. The first was that farmers no matter how "large" would want to be included in "small." Besides that, it was unclear whether the metric should be total acreage, gross income, net income, or something else and whether it should differ from state to state and from crop to crop. And then the question of what to count as a "farm" would have to be resolved. Would it be possible to combine all farm-based earnings per operator? Or would "large" farmers divide their operations among family members and into various family corporations and partnerships to qualify as "small"?

The possibility that government subsidies could be eliminated entirely seemed highly unlikely to nearly all the farmers we interviewed. The reason was not that farmers themselves would insist on retaining subsidies. Or that they would go under without subsidies. It was rather that government was the true beneficiary of these programs. Government's stake was having a reason to collect information about farming and exercise some control over farmers' activities.

"Government will never get out of the farming business in my opinion. Never. They want their fingers on it." This was how one of the wheat-belt farmers we interviewed put it. In his view farming was so important to the general well-being of the society that government officials knew they had to keep tabs on it. "If they cut us off," he ventured, "they won't know anything. I'll guarantee it. Those farmers won't tell them anything. This way the government knows what's going on here."

Although most of the farmers we spoke with thought some government intervention was necessary, a few expressed extreme distrust of government programs. They were disgusted not only with specific farm support programs but with government in general. In their view public officials were making bad decisions that were leading the nation to ruin. Democrats maybe were worse than Republicans, but neither party's leaders seemed to be garnering much respect.

"They're just promoting globalism," one woman explained, warming to the topic of how government policies were squeezing farm families out of business and favoring large investors and corporations that were driving up the price of land. "They're supposed to be intelligent people," her husband added, "but those guys never learned what two and two are."

This couple acknowledged that farm subsidies were benefiting them personally, but they thought subsidized crop insurance made more sense than direct payments. They thought subsidies should be restricted to small farmers. "That way we could keep more people on the land," the man explained. They also thought the government should curb deficit spending and lower the national debt, but were unsure how to accomplish those goals.

Their concern about globalism bordered on the belief that public officials were somehow in league with the United Nations and through it were promoting a single worldwide government. However, this concern was less about the aims of public officials and more about their own dependence on foreign markets. Like other farmers, they knew that foreign markets affected prices. They were even more concerned that equipment formerly produced in the United States was now being imported.

"We bought a small tractor last fall," the man explained, "and it was made in Turkey." That got him thinking about the John Deere tractor he owned that he assumed was made in the United States. He discovered that some of the John Deere's replacement parts were made in India. Some others were made in Hungary. The engine on another tractor came from Brazil. Everything was becoming so intertwined, he feared, that it would be impossible to get parts if some country decided not to trade with the United States.

This couple's concerns were more extreme than those of most of our interviewees. The more typical view was that globalism for better or worse was a fact of life. It had been for a long time, but was becoming increasingly important. Trade agreements, cheap transportation, and rising demand in developing countries were all part of the story.

"There's no doubt that we're in a worldwide economy now," a wheat-belt farmer said. It was interesting, he thought, that wheat prices in his area were probably affected less by local weather than by weather in Australia and Argentina. "If they're having bad weather," he said, "we're going to have good prices."

Although they knew the international markets in which they functioned were also shaped by government policies, few of our interviewees claimed to understand much about these policies. That was true among the ones who had studied them in college and who held office in state and regional farm associations, but was hardly surprising given the complexity of international trade agreements.

The tacit implication of being involved in international markets was feeling small. Contacting one's broker by cell phone to sell or buy in futures commodities was far different from meeting with one's neighbors at the local livestock auction barn. That was the lonelier side of being independent. It cast a shadow on being one's own boss. There were always bigger players in the game. It was good to be rooted in the land. Having a place of one's own provided a semblance of control. There was security in knowing one's limits.[10]

As for government, its role was to help in small ways. The typical view was that some government intervention was probably inevitable, even necessary, but was best kept to a minimum. As much as they disliked market fluctuations, they favored open markets. They understood and appreciated the need for laws protecting the soil and quality of the nation's food, but at the same time resisted the idea of "some pea-brained EPA official," as one farmer put it, telling them what to do.

There was not much hope, even among the larger and more financially successful farmers, of significantly affecting the way markets for farm goods operated. Technological innovation, niche marketing, and efficiencies of scale were interesting enough to generate enthusiasm and yet incapable of altering farmers' fundamental position. They were small producers in a complex system of trading that had been international in scope for a long time and was becoming increasingly so. That much was understood. And despite all that, they were glad to be in farming. They considered themselves fortunate to work with their families and to live close to the land. It was a good life, after all.

AFTERWORD

No one has a greater stake in the health of the land or the productivity of the land or the health of the society than farmers.

—Truck farmer, male, age 59

When you talk bad about farmers, don't do it with your mouth full.

—Farm country bumper sticker

The farmers we spoke with were adamant that more needed to be done to educate the public about farm life and its contributions to society. They took blame for not doing a better job themselves of providing information to the public. They hoped that sitting for an interview as part of the process leading to this book would help in a small way.

"We've got to look for common ground," one farmer explained. "We do come together three times a day," he observed, referring to the fact that the public interacts with farmers every day at mealtime. "Our urban neighbors don't really know very much about us and we find fault with them for not understanding us, but we're just as ignorant about what their life is all about. We need to connect the dots."

"We've got to co-exist and find common ground" was how another farmer put it. Pointing out that farmers depend on a lot of nonfarming industries and that many nonfarming industries depend on agriculture, he thought it was critical for each side to appreciate the other side.

Implicit in this comment is the view that farm life is more important nationally than the public may understand. If the common view is that hardly anyone in America farms, and if that view is further rooted in an understanding of modern history emphasizing the rise of industries and cities and the decline of farms, then it is not surprising that farmers want to revise that view.

The message farmers hoped the public would hear is that farmers are hard-working, responsible citizens. It was odd that they imagined anyone would think otherwise, and yet they read the newspapers often enough to see that some journalists saw them only as freeloaders who took money from the government even though they did not need it.[1]

Another message they hoped would come across was that farmers are not the stupid bumpkins who stayed behind because they could find nothing better to do with their lives. They generally felt that people they knew already understood this. There was nevertheless a lingering sense of insecurity that they wanted to set aside. They emphasized the range of skills necessary to succeed at farming and the increasingly sophisticated levels of technological knowledge involved.[2]

As producers of food and fiber, they wanted the public to have a better understanding of how food and fiber were produced. They hoped consumers would not hold them accountable when food prices increased. They wanted people to know that hardly any of the price of a loaf of bread went back to the farmer and that milk prices had risen far less than the cost of feeding cows.

Some of the farmers we spoke with were helping in small ways to educate the public. One was part of a committee that hosted local baseball games at which tables were staffed showing how much wheat went into a loaf of bread and how much corn went into ethanol production. Another belonged to a group that hosted visitors who spent a day on local farms getting their hands dirty feeding cows and tending pigs. Yet another served on the board of a statewide organization that put together press releases about farm policies.

What they appreciated about the information collected for the present project was that it provided a small window into their daily lives and into the meanings and values they associated with farming. The interviews gave them an opportunity to tell some of the stories about their current families and the generations who had gone before them. They appreciated the chance to describe their lives and to discuss their values in ways that could not be reduced to sound bites or headlines.

These stories show that farm life is indeed something that farmers feel passionately about, that it is not simply what they were left with after failing to pursue something more interesting and rewarding in the city. Few of the narratives suggest that farming is necessarily a *better* way of achieving a fulfilling life for oneself and one's family, only that it is distinctive and for that reason possibly difficult for outsiders to understand or appreciate.

Although it is different, farm life appears in the stories farmers tell as essentially similar in many respects to the values that other Americans embrace. In demographic terms farmers are older than the average American, more likely to be married and have children, and more likely to be of white European descent. And yet the messages their stories convey are that they love their families like everybody else, work hard, value their independence, and want to feel that they are contributing to the common good.

The families who were farming on the largest scale illustrated the fact that farming today is radically different from how it was practiced in earlier days. One of the fifth-generation farmers who talked with us was a clear example. He currently farms more than twice as much land as his father did and more than ten times as many acres as his grandfather. During wheat harvest he operates four combines with 30-foot headers, two tractors with grain carts, and three eighteen-wheelers, and it still takes him two weeks to finish. Wheat is only part of his business. He also plants hundreds of acres of irrigated corn and soybeans and feeds several hundred head of cattle.

Farming on that scale is the kind that evokes criticism from the nonfarming public. Critics worry that it cannot be good for the land, spells trouble for farming communities, demonstrates the folly of farm subsidy programs, and undercuts the very meaning of family farming. The criticisms are not wholly unfounded. This farmer uses chemicals instead of plowing, acknowledges that the water supply for irrigation is diminishing, and seldom has time to relax.

But farming of that kind is a far cry from agriculture run by corporate interests whose executives and shareholders have never set foot on a farm. This farmer is as worried about that prospect as anyone. He still considers his operation a family farm. It is no different from an extended family that runs a successful restaurant or trucking company. He repairs all the machinery himself to keep his input costs down. Most of the machinery belonged to his father before his father died a few years ago. His eighty-five-year-old mother rents it to him in return for some of the income from the cattle. He farms in partnership with his son, who lives nearby. Three of his grandsons are in 4-H. His other sons and their wives take their vacations when they can help with harvest. The reason he and his wife enjoy farming is that they are carrying on a family tradition. They both say it is in their blood.

The story from small farmers we spoke with was of course quite different. It was hard for them to get more land and thus difficult to achieve the efficiencies of scale they would have liked. They felt squeezed by the markets they could not control, the rising costs of fuel and fertilizer, and competition from corporate agriculture. They knew they were falling behind.

And yet the conclusion they drew was rarely that laws should be passed to somehow return farming to how it had been done in the past. It was not their view that technological innovation was bad or that changing farm practices spelled doom for family farming. It was certainly not their view that small farmers should be pitied.

The farmers we spoke with also hoped that their interviews would serve in small ways to help young people who might be thinking about farming as a career. When asked directly what advice they would give, they underscored the fact that getting into farming these days is very difficult. It takes so much money to purchase the necessary equipment that starting from scratch is almost prohibitive. In addition land is scarce, and even gaining the necessary skills is difficult unless a person grows up farming.

They nevertheless saw opportunities for young people from farm backgrounds whose families might have sufficient resources to help them get started. Despite all the changes that have taken place, they felt that farming was still a good life. It was a good way to earn a modest living knowing that one was doing interesting work that made a genuine contribution to the common good.

There were even opportunities for young people whose families did not have the resources to get them started. These were opportunities to learn the basic skills through summer jobs, internships, and part- or full-time jobs. Some of the truck and dairy farmers we spoke with were providing such opportunities, and some of the younger farmers had gotten started by working as paid laborers or as sharecroppers for other farmers.

At the same time farmers' advice emphasizes that career decisions rest on considerations other than economic costs and benefits. Their own decisions to farm included considerations about how they felt about living close to parents, what they loved doing and could do well, and what they wanted for their children.

The value of knowing how farmers in various locations made these decisions is that a person's own unique family history is placed in a wider perspective. For every story of sons or daughters farming congenially with their parents there are stories of misunderstandings and conflicts. Different families found different ways of resolving those tensions. Similarly, farmers vary in how much they interact with neighbors and in how they think about the role of churches in their communities. One size does not fit all.

An implicit message that came through as farmers described their grown children's careers was that people who grow up loving farm life are finding creative ways to stay somewhat involved even if they do not farm. It was evident that careers in farm-related activities, such as working in agronomy or for agribusiness corporations, provided these opportunities in some cases, while in others the ties to farming continued by holding jobs in rural communities, such as in teaching or health-related professions, or by returning to small towns in mid-career in order to do hobby farming.

The larger point that emerged implicitly from these farmers' interviews is that farming has been—and remains—a valued way of life for those who engage in it full-time. It is a way of life not only because people who farm live in the country and do different things from day to day than those who live in suburbs and cities. It is a way of life because of carrying on a family tradition so natural and so compelling that it seems to be in one's blood.

Farmers' attachment to this way of life is surely similar to how a second- or third-generation baseball player feels about the game. It probably resembles the feeling members of a tight-knit religious or ethnic group have about the distinctive ties, traditions, and shared memories that link them with one another.

Deep attachments of this kind need not be romanticized. They are traditions that include memories of hardship as well as of rosy events. Nor are they static. Ways of life change, and part of what gives people who share them a sense of accomplishment is their ability to adapt and to innovate.

For anyone wanting to understand farming as a way of life, it is the day-to-day meanings that matter. These meanings inhere in the land, in working the land, in the smells associated with it, in how the sun rises across the fields, in working among family members and neighbors, and in exercising one's independence in making difficult business decisions.

The traditional meanings that have given farm life a special place in farm families' hearts are clearly being strained by changes in contemporary agriculture. But instead of abandoning their sense that farm values are of intrinsic worth, farmers have been finding new ways to define the meaning of those values. Family farming is good for families, farmers say, not because family members work together the way they did in previous generations but because living in the country and seeing crops grow is valuable. Rented land in far-flung fields is no longer meaningful the way inherited family land was, but a farmer can still feel attached to it by improving it with the latest machinery and technology.

Whether the deep meanings inherent in farming are sufficiently robust to withstand contemporary challenges remains to be seen. It worries farmers that neighbors behave like sharks in the water and that more of these neighbors are investors who have little understanding of farm life or are corporations. When so much depends on new technologies and efficiencies of scale, they have reason to worry.

The other message that comes through in these interviews is important to remember, though. This is the fact that farming is fundamentally local and thus inherently diverse. If there is a subculture shared to an extent by farmers in different parts of the country, this is a pattern of meanings and values that is also refracted through the local adaptations that each family has made. It reflects the distinctive opportunities and constraints that the land provides.

APPENDIX

The research for this book grew out of a previous study that focused on the residents of small towns. That study included some residents who farmed or who owned land or whose businesses depended heavily on farmers in the area, but was concerned with how townspeople related to their immediate neighbors and what they did to make the town seem like a community. The research dealt with community festivals, schools, churches, businesses, and main street improvement efforts, and it focused on residents' aspirations for their towns and their families.

Comments about towns' locations and residents' attitudes toward the surrounding land came up often enough that a logical next step in that line of research was to study farmers. Interviews with community leaders and with farmers who lived in small towns in that study provided an opportunity to explore the various topics that were on farmers' minds. Through those conversations and from talking with rural sociologists and reading the literature, it was possible to develop a tentative list of topics to be examined.

Farm life is well documented in studies by agricultural economists, historians, and rural sociologists. Those studies amply describe the history of farming in the United States and its current characteristics such as state-by-state or county-by-county variations in the size of farms, crop yields, net farm income, and the like. The present project was conceived of as a qualitative supplement to that information.

To find out why a farmer may be especially attached to a particular parcel of land, the best way is to talk to that farmer. And the best way to capture the nuances and variations in those understandings is to talk with several dozen or a hundred farmers and listen closely to what they say. This is the role of qualitative research.

Qualitative research fills the gaps that cannot easily be filled by quantitative research. It reveals the meanings and experiences behind the generalizations that arise from other studies or that exist simply in the public mind as commonsense generalizations. If farmers are generally regarded as people who value being their own boss, what does that mean? And how are those meanings changing as farming changes?

If a young person grows up on a farm and feels a special pull to become a farmer but also wonders if farming is possible, the statistical profile of who farms and how much they earn is useful information. It can also be helpful to hear farmers talking about what is good about farming and what is not so good. Chances are, those are the kinds of stories that matter within families.

The questions that interest cultural sociologists are ones that depend especially on paying close attention to what people say, how they say it, and the meanings they ascribe to their experiences. Questions about the meanings of farming are no different. These are questions that are particularly interesting to investigate against the backdrop of changes taking place in American agriculture.

Farming may be a good place to raise children and to enjoy close family relationships. At least that is one impression one gains from superficial conversations with farmers. But is that changing as farms become larger and the technology involved becomes more sophisticated and expensive?

Similar questions can be asked about farmers' relationship to the land. How do utilitarian logics fit with logics grounded in nostalgia and family history? Does it matter if farmers spend less time on family land and more on rented acreage? How do they relate traditional understandings of the land to new relationships shaped by larger equipment and new technology?

Spending time with farmers, attending farm meetings, and listening in as they chat with neighbors over coffee, one can attain some answers to these questions. That is an excellent way of gaining an inside glimpse of the public culture that exists in a particular location. The literature includes several studies that do this well.

In-depth qualitative interviews provide a different kind of information. Many of the thoughts, interpretations, values, and stories that give meaning to farmers' day-to-day activities are not ones they discuss at farm meetings or with neighbors over coffee. This is not only because farmers, by their own admission, are rather tight-lipped about what they think and believe. It is true for other people as well. Musings about the daily meanings of life are personal.[1]

These meanings are fundamentally biographical. How they are expressed may be shaped by the particular social situation of the moment, but they extend beyond those moments. Biography implies history, continuity, and the location of particular events in larger narratives. These are the narratives that reflect and give expression to our outlooks on life.

In designing a study that would capture a range of stories, meanings, and opinions, I wanted to include farmers from several different regions, farmers who earned their living from different crops or livestock or other farm outputs, and farmers who were in the high, middle, or low ranges in terms of their overall scale of operations. I developed a purposive research design that would ensure inclusion of farmers within these various categories.

The goal was to talk with farmers who engaged in various kinds of farming but who otherwise were fairly typical of farmers in their communities. Other studies have examined special populations who in various ways are atypical, such as the Amish, African American sharecroppers in the South, single women who farm, and former residents of cities who are part of the back-to-the-land movement or who farm communally. The aim of the present study was to focus on ordinary farmers who earned their primary income from farming, considered farming their principal occupation, actually lived on a farm, grew crops or raised livestock in conventional ways, and yet varied in location and scale of operation.

The 2007 USDA Census of Agriculture (the most recent data available when my research began) reported that 347,760 US farms grew corn for grain, 84,317 grew corn for silage, 160,810 grew wheat, 18,605 raised cotton, 69,890 had dairy cows, 115,935 grew vegetables, and 69,172 had orchards. The states with the largest number of farms growing corn were Iowa, Illinois, and Minnesota; Kansas had the largest number of farms growing wheat; Texas and Georgia had the largest number of farms raising cotton; Wisconsin, Pennsylvania, and New York had the largest number of dairy farms; and California, Texas, Florida, and Washington had the largest number of farms growing vegetables and fruit.

I selected farmers for the study by first identifying four regions of the United States in which farming was characterized by different primary outputs. The first region was populated by farms primarily devoted to corn and soybeans, the second by farms raising wheat, the third by farms growing cotton, and the fourth by dairy and truck farming. Many of the farms in each region also included livestock and some specialized in other crops, such as alfalfa hay, milo, and rice. I then used the 2007 USDA Census of Agriculture data and selected three agricultural counties within each region, one of which was among the highest in its state in terms of overall value of agricultural output, one that was average, and one that was below average, and further determining that the counties selected were not contiguous and also varied in terms of average farm size.

The three corn-belt counties selected included a total of 1.6 million acres of farmland, 976,000 acres of which were planted in corn. The 2,500 farms in these counties included 1,520 principal operators whose primary occupations were farming and produced total agricultural output in 2007 valued at $756 million, of which $384 million was from corn. Variation among the three counties as measured by the ratio of standard deviation to mean was 0.26 for value of farm products sold and 0.77 for average farm size in acreage.

The three wheat-belt counties included 1.2 million acres of farmland, of which 310,000 were planted in wheat. The 1,700 farms included 810 principal operators whose primary occupations were farming and produced total agricultural output in 2007 valued at $946 million, of which $162 million was from wheat and most of the remainder was from livestock. The ratio of standard deviation to mean for the three counties was 0.41 for value of farm products sold and 0.75 for average farm size.

The three cotton-belt counties included 1.1 million acres in farming, of which 230,000 were planted in cotton. These counties included 1,400 farms, with 720 principal operators whose primary occupation was farming, and produced $292 million in agricultural output, of which $116 million was from cotton. The measures of variation were 0.28 for farm products sold and 0.81 for average farm size.

The three truck and dairy counties were composed of 1.9 million acres of farmland, of which 285,000 acres were in pastureland, vegetables, or orchards. They produced $1.7 billion of total agricultural output, of which $694 million was from dairy, vegetables, or fruit. There were 4,600 farms, with 2,450 principal operators whose primary occupation was farming. Variation among the counties was 0.45 for farm products sold and 0.43 for average farm size.

Within each cluster of counties, I initially identified individual farmers from publicly available information compiled by the Environmental Working Group, which lists the names of farmers in each county who have been the recipients of farm subsidies and the amounts they may have received through various government programs, such as commodity subsidies, crop insurance, and disaster programs. I stratified these names into four categories, ranging from those who had received the highest, higher than average, lower than average, and lowest amounts. I then selected names at random within each category, looked up the addresses of persons selected, and eliminated addresses that were in towns or cities.

With the assistance of the researchers working with me, we then sent a letter to each farm family that had been selected explaining the study and followed the letter by telephone. To compensate for potential biases in the lists obtained in these ways, we identified additional names through a snowball or referral method in which we asked the persons interviewed for recommendations of other farmers in their community. In each community we also interviewed community leaders, such as agricultural extension agents, farmers' cooperative managers, and clergy, and asked them for recommendations of farmers in the area.

We conducted 250 interviews in all. Of this number, 200 were with farmers and 50 were with community leaders in farming areas. Two-thirds of the interviews were concentrated in the twelve target counties, and the remaining third were in neighboring counties as a result of referrals and farmers having land in several counties. In all, the respondents were located in eight states. Twenty-five percent of the interviews were in areas in which corn and soybeans were the principal commodities, 28 percent were in predominantly wheat-growing areas, 13 percent were in cotton-growing areas, and 34 percent were in dairy and truck farming areas.

Besides these principal outputs, the farmers we interviewed were involved in a wide variety of secondary activities. The secondary outputs they produced included alfalfa hay, apples, asparagus, barley, beets, canola, carrots, catfish, cherries, edible organic cactus, hops, milo, mint, peas, pumpkins, rice, rye, strawberries, string beans, sugar beets, sunflowers, table grapes, and wine. Several were involved in specialized activities such as breeding cattle, repairing and selling farm equipment, and marketing hybrid seed grain. Thirty-two percent of the small grain farmers also raised livestock.

Of the farmers interviewed, the median age was 59, and when farmers who said they were still active but semiretired were excluded, the median age was 58, or approximately two years older than the median age of farm proprietors in national data. The oldest interviewee was 89, and the youngest was 21. Of the full-time farmers, 36 percent were in their sixties, 32 percent were in their fifties, 21 percent were in their forties, and 11 percent were in their twenties or thirties.

The median number of acres farmed, including land that was rented as well as land that was owned, was 1,450. The smallest acreage farmed was 10 and the largest was 16,000, not counting a corporate farmer with 35,000 acres, another with 50,000 acres, and a rancher with 64,000 acres. Twenty percent farmed fewer than 300 acres, and 35 percent farmed more than 2,500. When asked how their scale of operations compared with those of other farmers in their area, about half described it as average, a quarter said it was above average, and a quarter said it was smaller than average.

The median number of acres rented or owned by farmers interviewed in the corn-belt counties was 1,300, the smallest was 60 acres, and the largest was 12,000 acres. In the wheat belt, the median acreage among farmers interviewed was 3,000, the smallest was 400, and the largest was 16,000. Among the cotton growers, the median acreage was 3,000, the smallest was 200, and the largest was 7,000. Median acreage among the truck and dairy farmers was 700, the smallest was 10, and the largest was 8,500.

Twenty-one of the farmers interviewed individually and nine of the community leaders were women. The rest were men. Thirty-seven of the interviews were conducted jointly with husbands and wives. Thirty-eight percent of the farmers were fourth- or fifth-generation farmers, 57 percent were third-generation farmers, and 5 percent were first- or second-generation farmers. Twenty-six percent of the men in farming had no college training, 24 percent had some, and 50 percent were college graduates. Twenty-three percent of the women in farming had no college, 25 percent had some, and

52 percent were college graduates. Fifty-two percent of the women held full or part-time off-farm jobs. All but five of the farmers were currently married (two were divorced and three were widowed).

The interviews were conducted in person or by telephone by a team of ten trained interviewers. Several of the interviewers lived on farms or had lived on farms in the past and were intimately acquainted with farming in the areas in which they conducted interviews. The interviews followed a semistructured design that included standard questions asked verbatim and then followed by open-ended questions and additional probes to elicit further responses.

The interviews ranged in length from approximately an hour to more than three hours and took an hour and a half on average. Interviewees were informed that their names and the names of their counties would not be revealed. They were asked to speak candidly and tell their stories and express their opinions. The topics included questions about their family history in farming, reasons for choosing to farm, current farming activities, relationships with neighbors, religion, the land, technology, and markets.

The decision to withhold names and locations was made after experimenting with disclosing names and locations in a previous project that was largely about the historical development of particular communities. In that project it was necessary to identify the communities and to gain permission from individuals to reveal their names. It was evident, though, that knowing that their names would be used influenced what interviewees said. For example, they tended to put their communities in the best light possible and were reluctant to talk about problems and conflicts.

Withholding names and locations proved to be of considerable value in our interviews for the current project. Many of the farmers we spoke with noted as they responded to particular questions that these were comments they would not feel comfortable sharing at the coffee shop or if their names were being disclosed. They spoke openly about family relationships, conflicts, landlords, neighbors, farming decisions, and finances as a result.[2]

To accommodate farmers' seasonal schedules, we conducted interviews over a four-year period starting in 2010. The years were generally good in terms of yields and prices for corn, soybeans, wheat, and cotton, but were less good for dairy, livestock, and truck farming, and included periods of extreme drought or excessive rainfall and flooding in several of the locations in which we conducted interviews. In keeping with the logic of a purposive design, we revised the initial interview guide midway through the project to delete a few questions about which we had attained ample information and included some new questions that probed for additional information about some of the topics that emerged in the initial interviews.[3]

All the interviews were professionally transcribed and copies of the transcripts were sent to interviewees who wanted to have copies. The transcriptions yielded more than 5,000 pages of single-spaced text that served as the primary data for the project. The transcripts were analyzed qualitatively, reviewing the responses to all questions in each individual transcript and then comparing the responses to separate questions and topics across transcripts.

The interview data was supplemented with evidence from USDA Surveys of Agriculture, US Census and Bureau of Economic Analysis county-level data, and published and online sources about farming in the communities in which the interviews were conducted.

I am especially grateful to the farmers and community leaders in farming areas who took time from busy schedules to talk at length about their family histories in farming, what their daily lives are like, what they enjoy and do not enjoy about farming, and how they view the changing technologies and market conditions they are experiencing. I am indebted to everyone who participated in the research in these ways. Without their input, this book could not have been written.

The research process was truly a collective endeavor. Those who assisted at various points during the four years in which we conducted interviews and made sense of the information included Janice Derstine, Sylvia Kundrats, Karenna Martin, Paul Martin, Christi Martone, Karen Myers, Steve Myers, Cynthia Reynolds, Shayne Runnion, and Devany Schulz. They are the "we" to whom frequent reference is made in describing the research.

The work of scholars too numerous to mention contributed important ideas and insights along the way. Special appreciation for encouragement and inspiration over the years goes to Marvin Bressler, Paul DiMaggio, David Dobkin, Mitch Duneier, Christopher Eisgruber, Tom Espenshade, Marie Griffith, Katherine Rohrer, Jeffrey Stout, Shirley Tilghman, and Viviana Zelizer. The research also benefited from ideas sparked by countless editorials, essays in farm publications, news stories, and postings on the Internet.

Financial support for the research came from the Woodrow Wilson School of International and Public Affairs, the University Center for Human Values, and the Office of the Provost at Princeton University. Donna Defrancisco characteristically did a spectacular job of keeping the paperwork and bookkeeping straight.

I owe special thanks to my community of real and imagined rural conversation partners, among whom are Clarence Bain, Martin Bohnenblust, Arthur Dobrinski, Sarah Gutsch, Henry Keller, George Kratzer, Robert McCrory, Eldo McFarland, Marlin McFarland, Roland Rolfs, Violet Rolfs, Porter Skyles, Bill Teeter, Max Teeter, Louis Wilkey, Elmer Wuthnow, Evelyn Wuthnow, Kathryn Wuthnow, and Victor Wuthnow.

NOTES

INTRODUCTION

1. US Census Bureau, *Statistical Abstract of the United States: 2012* (Washington, DC: Government Printing Office, 2012), Table 616, "Employed Civilians, by Occupation, Sex, Race, and Hispanic Origin," based on the annual average of monthly figures from the Current Population Survey, civilian noninstitutional population 16 years old and over, occupational classifications as used in the 2000 census: 713,000 farmers and ranchers; 237,000 farm, ranch, and other agricultural managers. Estimates of the number of farmers (owners and tenants) from US Census Public Use Micro Samples, electronic data files, for earlier decades are 5.96 million in 1900, 6.37 million in 1910, 6.61 million in 1920, 5.99 million in 1930, 5.17 million in 1940, 4.38 million in 1950, 2.95 million in 1960, 1.78 million in 1970, 1.47 million in 1980, 1.05 million in 1990, and 738,000 in 2000. These estimates are drawn from the Occupation 1950 variable, line 100; the data files are courtesy of Steven J. Ruggles, Trent Alexander, Katie Genadek, Ronald Goeken, Matthew B. Schroeder, and Matthew Sobek. Integrated Public Use Microdata Series: Version 5.0 [machine-readable database] (Minneapolis: University of Minnesota, 2010). In 2007 the US Department of Agriculture estimated that there were 2.2 million farms in the United States and that of the total number of principal operators of these farms, 1.21 million were not farmers and 993,881 were farmers; USDA, *Census of Agriculture*, 2007, electronic data file, courtesy of the Inter-University Consortium for Political and Social Research, University of Michigan. The USDA figure includes farmers counted more than once because of serving as principal operator for more than one farm.

2. Although farming is of interest in the broader scholarly literature for the reasons suggested, its treatment in introductory sociology courses

largely emphasizes the historical passage of agrarian societies. For example, an examination of eight recently published introductory sociology textbooks revealed that farmers and farming were mentioned on average in less than a single paragraph and usually in reference to the evolution of human societies at a time when farming replaced hunting and gathering and was then replaced by industry. In these brief references, farming was associated with manual labor in subsistence economies and contrasted with modern industrial and technological developments. In one of the few references to contemporary farming, Jeanne H. Ballantine and Keith A. Roberts, *Our Social World: Introduction to Sociology*, 3rd ed. (Thousand Oaks, CA: Pine Forge Press, 2012), 153, state, "Postindustrial societies feature high dependence on technology and information sharing. Few people live and work on farms." They do not acknowledge that contemporary farming is also highly dependent on technology and information sharing.

3. Farming has received relatively little attention in mainstream social science journals in recent years, but several articles and books provide overviews of particular value. Linda Lobao and Katherine Meyer, "The Great Agricultural Transition: Crisis, Change, and Social Consequences of Twentieth-Century US Farming," *Annual Review of Sociology* 27 (2001), 103–24, is an excellent review of the recent sociological literature through the end of the twentieth century, focusing especially on the declining numbers of farmers, the extent of marginal farming, and the role of household labor. The essays in David L. Brown and Louis E. Swanson, editors, *Challenges for Rural America in the Twenty-First Century* (University Park: Pennsylvania State University Press, 2003), cover a range of social and economic issues. Bruce L. Gardner, *American Agriculture in the Twentieth Century: How It Flourished and What It Cost* (Cambridge, MA: Harvard University Press, 2002), is a treasure-trove of statistical information covering most of the twentieth century. Michael Mayerfeld Bell, *Farming for Us All: Practical Agriculture and the Cultivation of Sustainability* (University Park: Pennsylvania State University Press, 2004) is a beautifully written, well-argued discussion of the relationships of farming to markets, communities, families, selves, and technological change based on ethnographic research among sixty farmers in Iowa. Katrina Fried, *American Farmer: The Heart of Our Country* (New York: Welcome Books, 2008), is an engaging work of photojournalism for a general audience. Statistical information and summary reports from censuses of agriculture are available on the USDA website (www.agcensus.usda.gov).

4. Historical studies of American farming have, if anything, been richer than studies of recent developments. David B. Danbom, *Born in the*

Country: A History of Rural America (Baltimore, MD: Johns Hopkins University Press, 2006), is an engaging history of farming in America from pre-Columbian times to the present, paying particular attention to the twentieth century. The Depression has inspired a large literature of interesting studies, including especially the readable journalistic treatment in Timothy Egan, *The Worst Hard Time* (New York: Houghton Mifflin, 2006), and the detailed accounts in Pamela Riney-Kehrberg, *Rooted in Dust: Surviving Drought and Depression in Southwestern Kansas* (Lawrence: University Press of Kansas, 1994), and H. Craig Miner, *Next Year Country: Dust to Dust in Western Kansas, 1890–1940* (Lawrence: University Press of Kansas, 2006). The sections of Paul K. Conkin, *A Revolution Down on the Farm: The Transformation of American Agriculture since 1929* (Lexington: University Press of Kentucky, 2008), dealing with Depression-era farming in Tennessee are especially interesting. He also provides a brief summary of major twenty-first-century developments and challenges. My *Remaking the Heartland: Middle America since the 1950s* (Princeton, NJ: Princeton University Press, 2010) includes a discussion of the early settlement, agricultural history, and post–World War II developments in farming in nine Middle West states.

5. Conflicting or what might be regarded as contradictory social characteristics such as these are treated in various ways in the sociological literature, ranging from ignoring the cultural ways in which they are negotiated to conceiving of trade-offs between associated incommensurate values, investigating the proportions of a population that opts for various resolutions, or emphasizing the situational cues that result in the deployment of different cultural tool kits. The trade-offs and proportionality approaches impose forced choices that are unlikely to be present in real life, while the situational approach pays insufficient attention to underlying values and continuities. I follow a different approach that emphasizes the role that narratives and idioms play in both acknowledging and making sense of the cultural tensions that arise in conflicting, contradictory, and changing social contexts.

CHAPTER 1. FAMILIES

1. The role of separation of business enterprises from households in the development of modern rational capitalism is of course notably emphasized in Max Weber, *Economy and Society*, trans. by Guenther Roth (Berkeley: University of California Press, 1978), 375–80.

2. A prominent example of these arguments is the case of trust based on family networks in New York's diamond district given by James S. Coleman, *Foundations of Social Theory* (Cambridge, MA: Harvard

University Press, 1990), 109–13, an argument disputed in Eric Lesser, *Knowledge and Social Capital: Foundations and Applications* (Woburn, MA: Butterworth-Heinemann, 2000), 187; another argument focuses on the role of family networks in the success of business enterprises among immigrants—for example, Alejandro Portes, *Economic Sociology: A Systematic Inquiry* (Princeton, NJ: Princeton University Press, 2010).

3. Sonya Salamon, *Prairie Patrimony: Family, Farming, and Community in the Midwest* (Chapel Hill: University of North Carolina Press, 1992) is, in my view, one of the most insightful studies of farm families and one that provides rich information about the gender roles and intergenerational relationships in the farming communities she studied. Women's roles among my interviewees reveal shifts toward more complex understandings of partnership similar to those identified in Sarah S. Beach, "'Tractorettes' or Partners? Farmers' Views on Women in Kansas Farming Households," *Rural Sociology* 78 (2013), 210–28. Antecedents of current father-to-son farm succession and gendered division of labor in farm households are discussed in Rachel Rosenfeld, *Farm Women: Work, Farm, and Family in the United States* (Chapel Hill: University of North Carolina Press, 1987), and Jenny Barker Devine, *On Behalf of the Family Farm: Iowa Farm Women's Activism since 1945* (Iowa City: University of Iowa Press, 2013). Pamela Riney-Kehrberg, *Childhood on the Farm: Work, Play, and Coming of Age in the Midwest* (Lawrence: University Press of Kansas, 2005), discusses gender and birth order effects among farm children.

4. Using the cumulative data file of national surveys conducted between 1972 and 2012 makes it possible to identify nearly a thousand adults who list their current occupation as farming and to identify the occupations of those farmers' fathers. Eighty percent of the farmers' fathers were farmers; electronic data files, General Social Surveys, National Opinion Research Center, University of Chicago.

5. Memoirs and short essays by farmers and writers who grew up on farms are a rich source of family legends and traditions. Chronicling farm life mainly from the 1920s through the 1970s, this literature also provides background for understanding more recent similarities and differences. Carol Bly, *Letters from the Country* (Minneapolis: University of Minnesota Press, 1981), is set in Minnesota. Carol Bodensteiner, *Growing Up Country: Memories of an Iowa Farm Girl* (Des Moines, IA: Sun Rising Press, 2008), offers the perspective of being raised on a dairy farm. Howard Kohn, *The Last Farmer: An American Memoir* (Lincoln, NE: Bison Books, 2004), is located in Michigan from World War II to the 1970s. Carrie A. Meyer, *Days on the Family Farm: From the Golden Age through the Great Depression*

(Minneapolis: University of Minnesota Press, 2007), chronicles life on a farm in the Midwest during the first half of the twentieth century. Lawrence Svobida, *Farming the Dust Bowl: A First-Hand Account from Kansas* (Lawrence: University Press of Kansas, 1986), is a moving account originally written in the 1930s. Arnold J. Bauer, *Time's Shadow: Remembering a Family Farm in Kansas* (Lawrence: University Press of Kansas, 2012), traces the author's family from the 1930s through the end of the century. Rosemary Coplin Dahlberg's *Gravel and Grit: Childhood Memories of Life on a Kansas Farm* (Bloomington, IN: Crossbooks, 2010) focuses on the 1940s.

6. Of the farmers we spoke with, 51 percent of the men and 49 percent of the women were college graduates, a higher proportion than among farmers nationally, among whom 37 percent of men and 44 percent of women had completed some college education, according to the 2000 US Census; my analysis of the 5 percent public use electronic data file; courtesy of Steven J. Ruggles, Trent Alexander, Katie Genadek, Ronald Goeken, Matthew B. Schroeder, and Matthew Sobek. Integrated Public Use Microdata Series: Version 5.0 [machine-readable database]. Minneapolis: University of Minnesota, 2010.

7. Fifty-eight percent of the farmers and farm couples we interviewed described themselves as husband-wife farms, and a majority of these indicated some part-time or seasonal help that would have qualified them as a husband-wife-plus arrangement; 3 percent were family corporations; 12 percent involved formal partnerships; and 27 percent were informal partnerships. Nearly all of the formal and informal partnerships involved fathers and sons. Several of the partnerships were with brothers, one was with a sister, two involved fathers and daughters, and one involved a mother and two sons.

8. The complexity of family relationships that do not involve official partnerships poses interesting questions for interpreting statistics about principal operators, off-farm work, and farm size. Consider the arrangement a wheat-belt farmer in his mid-fifties described. He farms 5,000 acres, of which he owns 460. His son, who is in his late twenties, farms with him, but not in an official partnership. The son farms 160 acres on his own, but works for his father, receiving an hourly wage some of the time and free use of his father's machinery as payment-in-kind the rest of the time. Officially, each would be a principal operator and one would be classified as a large farmer and the other as a small farmer, but the classification would miss the extent to which they are working together. Additional complexity was evident in our interviews with farmers who were the principal operator of several different farms because of different landlords and rental agreements. For example, one farmer said he farmed three

farms, another said he farmed eleven, and yet another said he farmed thirty.

9. Two interesting studies of conflict in farm families are Fiona Gill, "Moving to the 'Big' House: Power and Accommodation in Inter-Generational Farming Families," *Rural Society* 18 (2008), 83–94, and John Hutson, "Fathers and Sons: Family Farms, Family Businesses, and the Farming Industry," *Sociology* 21 (1987), 215–29. Jane Smiley, *A Thousand Acres* (New York: Ballantine, 1991), is of course a powerful fictional rendition of conflict in farm families.

10. I have in mind the interesting work on money and intimacy in Viviana Zelizer, *Pricing the Priceless Child: The Changing Value of Children* (New York: Basic Books, 1985), and *The Purchase of Intimacy* (Princeton, NJ: Princeton University Press, 2005).

11. Ann R. Tickamyer and Debra A. Henderson, "Rural Women: New Roles for the New Century?" in *Challenges for Rural America in the Twenty-First Century*, edited by David L. Brown and Louis E. Swanson (University Park: Pennsylvania State University Press, 2003), 109–17, and Linda Lobao, "Gendered Places and Place-Based Gender Identities," in *Country Boys: Masculinity and Rural Life*, edited by Hugh Campbell, Michael Mayerfeld Bell, and Margaret Finney (University Park: Pennsylvania State University Press, 2006), 267–76, provide excellent reviews of the rural sociology literature on gender in farming communities. Sarah S. Beach, "'Tractorettes' or Partners? Farmers' Views on Women in Kansas Farming Households," *Rural Sociology* 78 (2013), 210–28, offers further evidence of changing understandings of gender roles.

12. It was possible to ascertain from the women or their husbands in nearly all cases whether the women held off-farm jobs. Overall, 58 percent did and 42 percent did not. The clearest dividing line was between women born during he 1940s or before and those born since 1950. Of the former, 42 percent held off-farm jobs; of the latter, 60 percent did.

13. Gender differences in geographic mobility are evident in data collected by the US Census in 2000, in which 34 percent of women who listed their primary occupation as farmer were living in a state other than their birth state, compared with 20 percent of men who did; my analysis of the 5 percent public use electronic data file.

14. I am especially grateful to Devany Schulz for the conversations we had and the insights she contributed through her junior independent paper at Princeton on the topic of conflict between daughters-in-law and mothers-in-law in farming communities. One of the few published studies of these relationships is Ramona Marotz-Baden and Claudia Mattheis, "Daughters-in-Law and Stress in Two-Generation Farm Families," *Family Relations* 43 (1994), 132–37.

15. Studies of farm succession that devote attention to the family dynamics involved include Norah C. Keating and Brenda Munro, "Transferring the Family Farm: Process and Implications," *Family Relations* 38 (1989), 215–19; Ayal Kimhi and Ramon Lopez, "A Note on Farmers' Retirement and Succession Considerations: Evidence from a Household Survey," *Journal of Agricultural Economics* 50 (1999), 154–62; and Shoshanah Inwood, Jill K. Clark, and Molly Bean, "The Differing Values of Multigeneration and First-Generation Farmers: Their Influence on the Structure of Agriculture at the Rural-Urban Interface," *Rural Sociology* 78 (2013), 346–70.

CHAPTER 2. NEIGHBORS

1. The argument that community is declining has been advanced most notably in Robert D. Putnam, *Bowling Alone: The Collapse and Revival of American Community* (New York: Simon and Schuster, 2000). A counter-argument is presented in Claude S. Fischer, *Still Connected: Family and Friends in America since 1970* (New York: Russell Sage Foundation, 2011). An interesting historical study showing the times and circumstances under which Americans have felt, often wrongly, that community was declining is Thomas Bender, *Community and Social Change in America* (New Brunswick, NJ: Rutgers University Press, 1978).

2. The challenges facing rural communities are amply described in David L. Brown and Kai A. Schafft, *Rural People and Communities in the 21st Century: Resilience and Transformation* (Malden, MA: Polity Press, 2011), and Richard Wood, *Survival of Rural America: Small Victories and Bitter Harvests* (Lawrence: University Press of Kansas, 2008).

3. Although the sociological literature emphasizes social capital and networks as the bedrock of community relationships, the approach taken in the sociology of culture also pays particular attention to shared narratives, storytelling and gossip, and the embedding of relationships in common myths and traditions—for example, see Donna Eder, "Cohesion through Collaborative Narration," *Social Psychology Quarterly* 51 (1988), 225–35; Donna Eder, "The Structure of Gossip: Opportunities and Constraints on Collective Expression among Adolescents," *American Sociological Review* 56 (1991), 494–508; and especially Suzanne Keller, *Community: Pursuing the Dream, Living the Reality* (Princeton, NJ: Princeton University Press, 2003).

4. Research on community attachment demonstrates that perceived neighboring is as important as actual friendship ties—for example, see Daniel R. Sundblad and Stephen G. Sapp, "The Persistence of Neighboring as a Determinant of Community Attachment: A Community

Field Perspective," *Rural Sociology* 76 (2011), 511–34; the farmers we spoke with and small-town residents we interviewed in a previous project suggest that perceived neighboring is rooted especially in instances of special caring that provide grist for community storytelling.

5. I examined this larger pattern of more geographically dispersed networks and the related development of community organizations adapting to open networks in *Loose Connections: Joining Together in America's Fragmented Communities* (Cambridge, MA: Harvard University Press, 1998).

6. The literature on social capital and neighborly social networks has been surprisingly silent about the particular norms that prevail when neighbors and friends are also competitors. People who live in proximity of one another in suburbs or small towns usually are not one another's competitors. If anything, neighborly relations are facilitated by mutual interest. Everyone benefits if neighbors keep up their lawn, participate in homeowners' associations, and keep an eye out for crime. But in other instances, people who work in close proximity to one another and are expected to get along and possibly be friends are actually engaged in a zero-sum game. In academic departments, for example, salaries are often subject to a fixed total amount, such that one faculty member receiving more means a colleague receiving less. In other situations, one person being promoted or receiving a bonus or an award means that a co-worker does not. Farmers' relationships with neighbors are not quite this competitive. But the infrequency with which land becomes available for rent or purchase means that one's neighbors are always one's competitors to an important extent.

7. Steven A. Camarota, *Immigrants in the United States, 2010: A Profile of America's Foreign-Born Population* (Washington, DC: Center for Immigration Studies, 2012), available at www.cis.org.

8. Robert Wuthnow, *Remaking the Heartland: Middle America since the 1950s* (Princeton, NJ: Princeton University Press, 2010). Tomas Roberto Jimenez, "Mexican-Immigrant Replenishment and the Continuing Significance of Ethnicity and Race," *American Journal of Sociology* 113 (2008), 1527–67, examines the impact of immigration on Garden City, a farming and meat processing town in southwestern Kansas.

9. Daniel T. Lichter, "Immigration and the New Racial Diversity in Rural America," *Rural Sociology* 77 (2012), 3–35, reviews the opportunities and challenges associated with immigration in farming communities and underscores the extent to which ethnic balkanization is taking place.

10. The 2010 US Census showed that the proportion of the population that was Hispanic in the towns located in the three corn-belt

counties in which our primary interviews were conducted averaged 1.4 percent; in the three wheat-belt counties, 5.3 percent; in the three cotton-belt counties, 10.3 percent; and in the three truck and dairy counties, 16.9 percent.

11. Edmund de Schweinitz Brunner, *Rural America and the Extension Service: A History and Critique of the Cooperative Agricultural and Home Economics Extension Service* (New York: Columbia University Press, 1949).

12. For interesting material on the Internet as it was beginning to be used in farming communities, see Michael J. Stern and Barry Wellman, "Rural and Urban Differences in the Internet Society: Real and Relatively Important," *American Behavioral Scientist* 53 (2010), 1251–56, and Florian Diekmann, Brian E. Roe, and Marvin T. Batte, "Tractors on eBay: Differences between Internet and In-Person Auctions," *American Journal of Agricultural Economics* 90 (2008), 306–20.

13. There were 131 towns in the 12 counties in which our primary interviews were conducted. The median population of these towns in 2010 was 1,176. The population in 62 towns was smaller in 2010 than it had been in 1980. The median population of these towns fell from 532 in 1980 to 445 in 2010. Fifty-six percent of the 42 towns in the three corn-belt counties lost population, 60 percent of the 15 towns in the wheat-belt counties did, 89 percent of the 19 towns in the cotton-belt counties did, and by comparison only 23 percent of the 56 towns in the dairy and truck counties did.

14. I examined population trends and perceptions of the effects of these trends among residents in *Small-Town America: Finding Community, Shaping the Future* (Princeton, NJ: Princeton University Press, 2013). One of the results presented in that volume was the extent to which population of towns in *primarily* farming areas declined. Using the USDA's Economic Research Service classification of county economic types, the median population of the 1,807 towns in "farming dependent" counties was 379 in 2010, and in 76 percent of the towns the population in 1980 had been larger than it was in 2010.

15. For an interesting ethnographic study of these changes, see Lyn C. MacGregor, *Habits of the Heartland: Small-Town Life in Modern America* (Ithaca, NY: Cornell University Press, 2010), and on the causes and effects of young people leaving, Patrick J. Carr and Maria J. Kefalas, *Hollowing Out the Middle: The Rural Brain Drain and What It Means for America* (Boston: Beacon Press, 2009). Other studies of rural and small town decline include Richard O. Davies, *Main Street Blues: The Decline of Small-Town America* (Columbus: Ohio State University Press, 1998), and Osha Gray Davidson, *Broken Heartland: The Rise of America's Rural Ghetto* (Iowa City: University of Iowa Press,

1996). Earlier decline in farming towns is described in numerous articles and books—for example, see H. Paul Douglass, *The Little Town: Especially in Its Rural Relationships* (New York: Macmillan, 1919); Richard Lingeman, *Small Town America: A Narrative History, 1620–the Present* (New York: Putnam, 1980); and Carroll Engelhardt, *The Farm at Holstein Dip: An Iowa Boyhood* (Iowa City: University of Iowa Press, 2012).

CHAPTER 3. FAITH

1. Roger Finke and Rodney Stark, "Religious Economies and Sacred Canopies: Religious Mobilization in American Cities, 1906," *American Sociological Review* 53 (1988), 41–49, is a helpful source that challenges the view that religion held a natural advantage in rural areas as opposed to urban areas at the start of the twentieth century. They challenge the idea that religion was weaker in urban areas because of greater diversity there that, according to some theories, would reduce the plausibility of any particular faith. They argue that diversity actually heightened competition and thus encouraged congregations to innovate and be more aggressive in seeking members. A larger account would need to include the fact that diversity and competition were also present during the latter half of the nineteenth century in most rural communities. The advantage of urban churches over rural churches, though, included shorter commuting distances from homes to church, more of the population employed in regular weekday jobs, fewer daily chores than on farms, electricity, and growing populations.

2. Farmers and farming communities can be studied without paying attention to religion—and usually are. However, the culture and values and meanings of farming cannot be. We found that religion inflected farmers' interpretations of their lives and that participation in religious organizations was so common in their communities that they either did participate or went to some pains to explain why they did not. As a practical matter, clergy who serve in farming communities are important community leaders who need to understand what farmers think and believe privately about religion as well as what they may say at church meetings. In the sociological literature, religion is understood to have been an important part of local culture in agrarian societies and for that reason is sometimes considered to be less important in modern industrial societies. How religion is understood among contemporary farmers is thus of broader importance in this literature.

3. Nor did they express any affinity with so-called New Age beliefs that drew from the historical traditions of American nature religion. These traditions are amply described in Catherine Albanese, *Nature*

Religion in America: From the Algonkian Indians to the New Age (Chicago: University of Chicago Press, 1990).

4. Data collected in 2000 as part of a national study of congregations and church membership showed that in counties classified by the USDA as "farming dependent," the median number of churches per county was 26 and the median number of adherents was 5,081. The rate of adherence was 80 percent of the total population in farming dependent counties with total populations of less than 5,000; 79 percent in counties with populations of 5,000 to 9,999; 73 percent in counties with populations of 10,000 to 19,999; and 67 percent in counties with populations of 20,000 to 49,999 (the figure for all US counties was 67 percent). In counties with populations of less than 5,000 the median number of churches was 15, yielding an average number of adherents per congregation of 160; in the larger counties the respective median number of congregations rose to 26, 39, and 56, respectively, and the respective median number of adherents per congregation rose to 204, 236, and 269, respectively. Between 1980 and 2000 the number of congregations remained nearly constant (declining only from 17 to 15 in the smallest counties), but the number of adherents fell by 19 percent in counties with total populations of less than 5,000, by 12 percent in counties with populations of 5,000 to 9,999, and by 4 percent in counties with populations of 10,000 to 19,999; there was a 10 percent increase in the counties with populations of 20,000 to 49,999. These figures are for farming-dependent counties only. Many of the nation's farmers live in counties that are economically diversified. Electronic data files for the religion data were produced by the Glenmary Research Center and made available through the Association for Religion Data Archives (www.thearda .com). Additional information and analysis about religion in small towns and rural areas is included in Wuthnow, *Small-Town America.*

5. For material focusing on the practical aspects of ministry in rural congregations, see Miriam Brown, editor, *Sustaining Heart in the Heartland: Exploring Rural Spirituality* (Mahwah, NJ: Paulist Press, 2005); Peter G. Bush and H. Christine O'Reilly, *Where 20 or 30 Are Gathered: Leading Worship in the Small Church* (Herndon, VA: Alban Institute, 2006); Lawrence W. Farris, *Dynamics of Small Town Ministry* (Herndon, VA: Alban Institute, 2000); and Shannon Jung, editor, *Rural Ministry: The Shape of the Renewal to Come* (Nashville: Abdingdon, 1998).

6. These observations were drawn from qualitative interviews with thirty clergy representing eleven different denominations or traditions and located in the twelve counties in which primary interviews were conducted with farmers. Average attendance at weekly worship services ranged from twenty to five hundred. The median was sixty. According

to InfoGroup data for 2010 available from www.Socialexplorer.com, there were 713 churches in these twelve counties, approximately 400 of which were located in rural communities or towns under 25,000 in population; inspection of map locations suggested that 106 congregations were country churches; in the InfoGroup data, 27 percent of church members in these counties were classified as evangelical Protestant, 36 percent as mainline Protestant, 27 percent as Roman Catholic, and 10 percent as "other," including Latter-Day Saints and Jehovah's Witnesses.

7. I am grateful to Mary Jo Neitz, "Reflections on Religion and Place: Rural Churches and American Religion," *Journal for the Scientific Study of Religion* 44 (2005), 243–47, and Mary Jo Neitz, "2008 Association for the Sociology of Religion Presidential Address: Encounters in the Heartland: What Studying Rural Churches Taught Me about Working across Differences," *Sociology of Religion* 70 (2009), 343–61, whose research on rural churches in Missouri has emphasized the variety among them and especially the fact that some of these churches are in exurban areas with growing populations and significant percentages of newcomers. That was evident among the pastors we spoke with in our research and in a previous study of churches in small towns.

CHAPTER 4. INDEPENDENCE

1. Alexis de Tocqueville, *Democracy in America* (New York: Harper & Row, 1966, originally published in 1835).
2. Robert N. Bellah, Richard Madsen, William M. Sullivan, Ann Swidler, and Steven M. Tipton, *Habits of the Heart: Individualism and Commitment in American Life* (Berkeley and Los Angeles: University of California Press, 1985).
3. For an overview of the literature on this topic, see Rebecca J. Erickson, "The Importance of Authenticity for Self and Society," *Symbolic Interaction* 18 (1995), 121–44.
4. The status aspect of farmers being more self-reliant than neighbors in other occupations is evident in the narratives included in Melissa Walker, *Southern Farmers and Their Stories: Memory and Meaning in Oral History* (Lexington: University Press of Kentucky, 2006).
5. The understanding of practice here is from Alasdair MacIntyre, *After Virtue: A Study in Moral Theory*, 2nd ed. (Notre Dame, IN: University of Notre Dame Press, 1984), and Jeffrey Stout, *Ethics after Babel: The Languages of Morals and Their Discontents* (Boston: Beacon Press, 1988); it emphasizes rule-governed behavior that is learned, grounded in social norms, and pursued in ways that include a sense of personal mastery.

6. Some of the other farmers we spoke with also made it clear that the reason they especially enjoyed the diversity of tasks and skills involved was the challenge it presented. They did not mean to suggest that farming was very different in this respect from many other occupations consisting of varied nonroutine work. It was rather the challenge of getting up in the morning, having decisions to make, and knowing that each day would likely be different from any other day. The independence they associated with that kind of work included a sense of personal freedom and the feeling that one's abilities were being put to full use.

7. The larger related point about American individualism involving well-institutionalized roles is developed in David John Frank and John W. Meyer, "The Profusion of Individual Roles and Identities in the Postwar Period," *Sociological Theory* 20 (2002), 86–105.

CHAPTER 5. THE LAND

1. Although much of the literature on farmland has dealt with its economic aspects, the sociological and anthropological literature includes studies examining the meanings of place and the emotional aspects of land ownership and cultivation—see for example, Michael Mayerfeld Bell, "The Ghosts of Place," *Theory and Society* 26 (1997), 813–36. Historical studies reveal the extent to which meanings of land changed as more of the population lived in cities and the nation shifted from an agrarian to an industrial economy—for example, see Edwin C. Hagenstein, Sara M. Gregg, and Brian Donahue, editors, *American Georgics: Writings on Farming, Culture, and the Land* (New Haven, CT: Yale University Press, 2011). Attention has also been devoted to the so-called new agrarian movement, in which the values of an older, simpler, smaller, healthier, and presumably more conservation-minded relationship to the land are emphasized—for example, see Eric T. Freyfogle, *The New Agrarianism: Land, Culture, and the Community of Life* (Washington, DC: Island Press, 2001). Less attention has been given to how the meanings of land among farmers themselves may be changing as a result of recent technological developments.

2. Scholars of American agriculture have expressed concerns especially about the effects of these changes on the values associated with family farming. As an example, Melissa Walker, "Contemporary Agrarianism: A Reality Check," *Agricultural History* 86 (2012), 1–25, writes, "I was raised by a man who believed farming was the highest human calling. I would love to see family farms flourish in this nation if only so that those who feel called to farm can do that work. Growing up, I was taught that life on the land made better people than life off it. Part

of me still embraces the potential of rural life to mold people of good character." She continues, "I believe that industrial agriculture has harmed our environment; industrial farming practices have contaminated soil and water and created monocultures that have increased, not reduced, the incidence of insect infestations and disease among plants and livestock" (p. 3).

3. Although a majority of the farmers we spoke with were at least third-generation farmers whose land included some that had been in the family for decades and was especially valued for that reason, it is also useful to note the extent to which geographic mobility has been observed historically as a characteristic of American farming. Alexis de Tocqueville wrote, "It seldom happens that an American farmer settles for good upon the land which he occupies; especially in the districts of the Far West, he brings land into tillage in order to sell it again, and not to farm it; he builds a farmhouse on the speculation that, as the state of the country will soon be changed by the increase of population, a good price may be obtained for it" (*Democracy in America*, 1835, chapter 19; available at xroads.virginia.edu). Studies showing extensive geographic mobility among farmers in the 1870s include Allan Bogue, *From Prairie to Corn Belt: Farming on the Illinois and Iowa Prairies in the Nineteenth Century* (Chicago: University of Chicago Press, 1963), and James C. Malin, "The Turnover of Farm Population in Kansas," in *History and Ecology*, edited by Robert P. Swierenga (Lincoln: University of Nebraska Press, 1984), 269–99, originally published in *Kansas Historical Quarterly* 4 (1935), 339–72. The farmers we spoke with generally knew something of the history that had resulted in ancestors settling where they did and staying, which added to the emotional value of land still in the family.

4. Their view of being at home with the land resembled that observed among adults who had grown up on farms in the study by Pamela Riney-Kehrberg, *Childhood on the Farm: Work, Play, and Coming of Age in the Midwest* (Lawrence: University Press of Kansas, 2005), 210–34.

5. The tensions between wanting to be at home and wanting to be free to roam are discussed in Susan J. Matt, *Homesickness: An American History* (Oxford: Oxford University Press, 2011).

6. Mr. Freeman's story came as close as any of the farmers we spoke with to resembling the back-to-the-land sentiments featured in Rebecca Kneale Gould, *At Home in Nature: Modern Homesteading and Spiritual Practice in America* (Berkeley: University of California Press, 2005).

7. With respect to the memories associated with barns, an interesting statistic reported in the USDA's 2007 census of agriculture was that 664,264 farms nationwide had a barn built prior to 1960 (USDA,

2007 Census of Agriculture, table 44, available at www.agcensus.usda
.gov). In the counties in which our primary interviews were conducted, there was an average of 297 farms per county with barns of
this vintage in the corn-belt counties, 201 in the wheat-belt counties, 83 in the cotton-belt counties, and 378 in the truck and dairy
counties.

8. Another wheat-belt farmer used almost identical language in saying
that the considers the land he farms as "more of a living organism than
just a piece of dirt."

9. Although the farmers we spoke with were perceiving the benefits of
understanding soil chemistry in new ways, farmers' interest in soil
chemistry is hardly new, as discussed in Benjamin R. Cohen, *Notes
from the Ground: Science, Soil, and Society in the American Countryside*
(New Haven, CT: Yale University Press, 2009).

CHAPTER 6. TECHNOLOGY

1. Examples of studies of the economic determinants of technological
adoption in agriculture include Franz Batz, Kurt J. Peters, and Willem
Janssen, "The Influence of Technology Characteristics on the Rate
and Speed of Adoption," *Agricultural Economics* 21 (1999), 121–30;
Gershon Feder and Dina L. Umali, "The Adoption of Agricultural
Innovations: A Review," *Technological Forecasting and Social Change*
43 (1993), 215–39; L. D. Hiebert, "Risk, Learning, and the Adoption of Fertilizer Responsive Seed Varieties," *American Journal of Agricultural Economics* 56 (1974), 764–68; and Sunil Thrikawala, Alfons
Weersink, Gary Kachanoski, and Glenn Fox, "Economic Feasibility
of Variable-Rate Technology for Nitrogen on Corn," *American Journal of Agricultural Economics* 81 (1999), 914–27.

2. We asked the farmers we interviewed, "Since you started farming,
what are the biggest changes you personally have experienced in the
way farming is done?" Eighty-six percent mentioned technological
changes, such as larger and more sophisticated machinery, genetically
engineered seed, and new herbicides and pesticides; the remaining
14 percent mentioned rising prices, larger farms, and more government regulations.

3. Or as a blogger who had recently moved from the city back to the
farming community in which he had been raised described them,
these tractors were like "day spas on giant wheels"; Matthew James,
"Tractors Drive Themselves: One Man's Return to the Farm," 2012,
available at www.mcsweeneys.net.

4. The USDA reported that 67 percent of farms had Internet access in
2013, up from 50 percent in 2002, and that 62 percent of those with

access reported having a high-speed connection; USDA *Farm Computer Usage and Ownership* (Washington, DC: National Agricultural Statistics Service, 2013), available at www.agcensus.usda.gov.

5. A standard bin of apples includes twenty-five 40-pound boxes, or a total of a thousand pounds and, depending on the size of apples, holds about two thousand apples.

6. USDA, "Adoption of Genetically Engineered Varieties of Corn, Upland Cotton, and Soybeans, by State and for the United States, 2000–12," July 2012, available at www.ers.usda.gov/data-products.

7. Rachel Schurman and William A. Munro, *Fighting for the Future of Food: Activists versus Agribusiness in the Struggle over Biotechnology* (Minneapolis: University of Minnesota Press, 2010).

8. On concerns about the risks of technological change for the environment and quality and safety of the food supply, popular treatments include Eric Schlosser, *Fast Food Nation: The Dark Side of the All-American Meal* (New York: Harper Perennial, 2001); Karl Weber, *Food, Inc.: How Industrial Food Is Making Us Sicker, Fatter and Poorer—and What You Can Do about It* (New York: Public Affairs, 2009); and Jill Richardson, *Recipe for America: Why Our Food System Is Broken and What We Can Do to Fix It* (Brooklyn, NY: Ig Publishing, 2009).

CHAPTER 7. MARKETS

1. Wayne S. Cole, *Roosevelt and the Isolationists, 1932–45* (Lincoln: University of Nebraska, 1983); Lawrence Goodwyn, *The Populist Movement: A Short History of the Agrarian Revolt in America* (New York: Oxford University Press, 1978); and Robert C. McMath, *American Populism: A Social History, 1877–1898* (New York: Hill and Wang, 1990).

2. In a different context, an illuminating study of the feelings of efficacy or inefficacy that come from being able to shape prices is Olav Velthuis, "Symbolic Meanings of Prices: Constructing the Value of Contemporary Art in Amsterdam and New York Galleries," *Theory and Society* 32 (2003), 181–215, and more fully developed in Olav Velthuis, *Talking Prices: Symbolic Meanings of Prices on the Market for Contemporary Art* (Princeton, NJ: Princeton University Press, 2007).

3. An interesting study suggests that agricultural intermediaries also exercise a role in dampening cooperation; Brent Hueth and Phillippe Marcoul, "Information Sharing and Oligopoly in Agricultural Markets: The Role of the Cooperative Bargaining Association," *American Journal of Agricultural Economics* 88 (2006), 866–81.

4. How farming is affected and in some cases replaced by large-scale agribusiness entities is examined with reference to the meat processing industry in Donald D. Stull and Michael J. Broadway, *Slaughterhouse Blues: The Meat and Poultry Industry in North America* (San Francisco: Wadsworth, 2004), and Wilson J. Warren, *Tied to the Great Packing Machine: The Midwest and Meatpacking* (Iowa City: University of Iowa Press, 2007). The history of two of the nation's largest agribusiness conglomerates is described in Wayne G. Broehl Jr., *Cargill: From Commodities to Customers* (Hanover, NH: University Press of New England, 2008); Brewster Kneen, *Invisible Giant: Cargill and Its Transnational Strategies*, 2nd ed. (New York: Pluto Press, 2002); and Lawrence Busch, William H. Friedland, Lourdes Gouveia, and Enzo Mingione, *From Columbus to ConAgra: The Globalization of Agriculture and Food* (Lawrence: University Press of Kansas, 1994).

5. The USDA's 2007 census of agriculture reported that 20,437 of the nation's 2.2 million farms (0.0093) were engaged in organic production; USDA, *2007 Census of Agriculture*, table 43, available at www.agcensus.gov. In the counties in which our primary interviews were conducted, organic farming was practiced at 133 farms in the truck and dairy counties but at only 18 farms in any of the other counties.

6. However, another source we talked with described organic cotton that did not have to be dyed because it was naturally the shade of denim.

7. Farmers markets selling items directly to consumers grew five- to tenfold in various states following passage of the 1976 Farmer-to-Consumer Direct Marketing Act; Allison Brown, "Counting Farmers Markets," *Geographical Review* 91 (2001), 655–74; Theresa Varner and Daniel Otto, "Factors Affecting Sales at Farmers' Markets: An Iowa Study," *Review of Agricultural Economics* 30 (2008), 176–89. On health insurance, see Ziaoyong Zheng and David M. Zimmer, "Farmers' Health Insurance and Access to Health Care," *American Journal of Agricultural Economics* 90 (2008), 267–79.

8. In several of the areas in which we conducted interviews, agricultural corporations were also restricted by state law to having only a certain relatively small number of owners and requiring that those owners all live within the state.

9. The perception that "farmers are greedy wards of the state," as a farm journalist observed with regret during discussion of the 1985 farm bill, had been around for a long time; Dan Looker, "A Proud Breed in a Hostile World," *Prairie Schooner* 60 (1986), 39–44, quote is on page 39. A less sympathetic treatment appears in Richard Manning, "Against the Grain: A Portrait of Industrial Agriculture as a Malign Force," *American Scholar* 73 (2004), 13–35, who writes, "I

have come to think of agriculture not as farming but as a dangerous and consuming beast of a social system" (p. 16).

10. The norms of camaraderie and rivalry at local auction barns are vividly described in George A. Boeck, *Texas Livestock Auctions: A Folklife Ethnography* (New York: AMS Press, 1989).

AFTERWORD

1. Between 2001 and 2009 the *New York Times*—significant not only as a leading national newspaper but also as the source of items frequently reprinted in local newspapers and covered on television—carried only 33 stories in which the phrase "farm families" appeared and the story was not about farm families in other countries. The stories focused on farm subsidies, the decline of farms and small towns, rural poverty, longing for the past, and occasional technological developments such as wind farms. During the same period the *New York Times* printed 362 stories in which the phrase "farm bill" appeared. These stories emphasized health concerns, soil contamination, subsidies, welfare, taxes, and conservatism, and included headlines with words and phrases such as "disgraceful," "misguided," "hypocrisy," "bias," "panderer," "nightmare," "farm belt follies," "farm aid," "illegal farm subsidies," "harvest of shame," "amber fields of bland," "politics as usual," "prices rise," "sweetheart deal," "failed frontier," "seeds of decline," and "forbidden fruits."

2. Judging from poll results, farmers' perception of being looked down on by the general public may be unfounded, despite negative portrayals in major newspapers. A Gallup Poll conducted in August 2012, for example, showed that 52 percent of respondents were "very positive" or "somewhat positive" toward "farming and agriculture," while only 19 percent were "very negative" or "somewhat negative." The 52 percent for farming was higher than for any other industries except retail, restaurants, and computers. In comparison, only 22 percent were positive toward the oil and gas industry, 23 percent were positive toward the federal government, and 25 percent were positive toward banking. Gallup Poll conducted August 9 to 12, 2012, based on 1,012 telephone interviews among a national sample of adults; courtesy of the Roper Center for Public Opinion Research, University of Connecticut, Storrs, CT.

APPENDIX

1. The logic and methods of qualitative interviewing are amply described in Steiner Kvale and Svend Brinkmann, *InterViews: Learning*

the Craft of Qualitative Research (London: Sage, 2009), and Irving Seidman, *Interviewing as Qualitative Research: A Guide for Researchers in Education and the Social Sciences* (New York: Teachers College Press, 2006). Mario Luis Small, "'How Many Cases Do I Need?' On Science and the Logic of Case Selection in Field-Based Research," *Ethnography* 10 (2009), 5–38, is a good source on the relationship between qualitative interviews and other qualitative methods. In a different context I have discussed the strengths and limitations of qualitative interviews in "Taking Talk Seriously: Religious Discourse as Social Practice," *Journal for the Scientific Study of Religion* 50 (2011), 1–21.

2. Joan M. Jensen, "Telling Stories: Keeping Secrets," *Agricultural History* 83 (2009), 437–45, provides background that further underscores the value of anonymity. In our interviews we noted a tendency in our pilot interviews for farmers and for farmwomen especially to give terse responses. We modified the interview guide accordingly by asking longer, chattier questions that created a more conversational atmosphere and succeeded in eliciting more detailed responses. We also prompted specifically for stories and interpretations by asking interviewees to take a few minutes to tell stories on particular topics.

3. On purposive design, see especially Kathy Charmaz, *Constructing Grounded Theory: A Practical Guide through Qualitative Analysis* (London: Sage, 2006), and Juliet Corbin and Anselm C. Strauss, *Basics of Qualitative Research: Techniques and Procedures for Developing Grounded Theory* (London: Sage, 2007).

INDEX

accidents, 4, 9, 21, 39–40, 72, 75, 92. *See also* hardships; injuries; risks

African Americans, 60

aging, 3, 43–45; and church participation, 81, 82; and faith, 73; and machinery, 146; and self-worth, 44. *See also* farmers, older; grandparents; retirement/retirees

agrarian societies, 199–200n2

agribusinesses, 114; career with, 188; competition from, 142, 169, 173; and market, 169; and seed, 134, 161, 173. *See also* corporations

agricultural extension services and agents, 33, 66, 67, 155, 167, 194

agricultural intermediaries, 214n3

agritourism, 63, 131

agronomists, 104, 105, 180; and technology, 150, 160, 161

agronomy, 28, 42, 45, 66; advances in, 136; and business decisions, 111; careers in, 188; and safety concerns, 137

ancestors, 29, 39; and adaptability, 170; betrayal of, 22; connection with, 14; and corporations, 178; and faith, 89, 93; and family tradition, 18; following in footsteps of, 2, 14, 18; hardships of, 19, 21; and land, 122, 123, 128; and legitimacy, 15; and risk, 172; stories about, 19, 212n3; and women, 36. *See also* families; grandparents

Archer Daniels Midland, 177

artificial insemination, 157

Asian Americans, 60

aunts, 24, 25, 179

authenticity, 40, 42, 98–99

bankers, 33, 58

banks, 19, 20, 69, 116; credit with, 59; loans from, 10, 21, 23, 25, 148. *See also* debt

baptisms, 90

Baptists, 77

bereavement, 73, 90. *See also* death

Bible, 74, 76, 77, 79, 86, 135

biodiesel, 158

biography, 192

booster clubs, 63

brothers, rivalries between, 28, 32. *See also* partnerships

BST (bovine somatotropin), 154

business, 8, 25; assets brought to, 31; and assistance to neighbors, 50; bookkeeping and paperwork in, 2, 24, 25, 26, 37, 40, 132, 137, 155; children's exposure to, 40; as conducted across generations, 25; and conflicts, 27–33; essential, 69; and family, 7, 8, 9, 13, 24–27, 29–30, 32, 108–9; and farm towns, 69; formal agreements in, 5, 25, 27, 32, 53, 203n7; and income taxes, 26, 27; informal agreements in, 25; and informal friendly relationships, 53; informal partnerships in, 203n7; and information technology, 150, 151–52; as kept to self, 116; and land, 121, 132; large, 69; local, 69; mature decisions in, 29; and partnerships between generations, 43; rational decision-making in, 7; and technology, 141; and transition agreements, 30; traveling for, 54; verbal and written agreements in, 26–27, 29–30, 31, 32. *See also* partnerships

capitalism, 93

Catholics, 75, 78, 83, 85, 87

cattle business, 5, 25, 150, 157, 195. *See also* dairy farmers

cell phones, 149, 151, 160, 161

Chamber of Commerce, 63

character, 20, 21, 29, 115

chemical companies, 69, 153–54, 155

chemicals, 120, 154, 187; ambivalence about, 156, 157, 160; cost of, 158; differing views of, 167; and government, 156, 157; knowledge about, 66, 67, 117; and land, 10, 154, 161; mistakes with, 154; and newcomers, 62; purchase of, 150; safety of, 154, 156; and social relationships, 155; and stewardship, 136, 137; and technology, 140, 141. *See also* fertilizer; herbicides; pesticides

children, 38–43; and ability to become farmers, 160; and authenticity, 40; and business decisions, 40; and changes in farm life, 40–41, 42; and choice of farming as career, 2, 17–19, 36, 42–43, 97, 100; chores for, 2, 13–14, 16, 17, 38, 41, 98; and churches, 81; and common sense, 41–42; conflicts with in-laws over, 36; and conflicts with parents, 9; and country living, 41; and education, 42; expectations for, 188; and farm equipment, 41; farms as good place to raise, 9, 38, 41, 42, 61, 142, 162, 192; in farm towns, 68–69; and farmwork, 24, 38–39, 40, 43; freedom of, 41; help for, 24, 25, 43, 44, 55–56; help for to start farming, 29, 160; and isolation, 42; and life on farm, 38, 98; maturity of, 41; memories of being, 8, 14, 16, 20, 38–39, 40–41, 43, 98, 99; and nature, 42; parents' absence from, 42; as raised in farm families, 14; and respect for parents, 40; and risks, 39–40; and sense of place, 39; skills and experiences attained by, 40–42, 98; sports and town activities for, 41, 43; and technology, 146–47, 160; and towns vs. farms, 41; values instilled in, 13, 42, 43; and women's roles, 34; and work ethic, 41, 42, 43

churches: activities of as practical, 82–83, 84; aging population of, 81, 82, 88–89, 90; attendance and participation at, 64, 65, 73, 74, 75, 80–86, 88, 92, 93–94; and community-wide crises, 84; conflicts in, 91–92; and declining population, 90; denominational boards of, 83; and economic assistance, 84; farmers and non-farmers in, 85, 86, 87; farming as more important than, 90–91; and food-sharing activities, 83–84; and illness, death, and tragedy, 83; and intermarriage between denominations,

85; interrelatedness of congregations in, 87; and joining and helping ethic, 82; and large farmers, 80–81; meaningful activities of, 90; and mission trips, 83; multigenerational loyalty to, 93; number of, 209n4; organizational structure provided by, 84; and prayer chains, 84; preaching and teaching in, 83; reports on in popular press, 80; in rural communities, 72; and shop owners, 87; small-town and country, 86; social role of, 73, 82, 83; and Sunday morning classes, 88; in town, 85; and traditions, 85, 89, 90; urban vs. rural, 208n1. *See also* clergy; neighbors; religion and faith

cities: bankers and investors from, 58; business contacts in, 68; choice of, 22, 23; and declining farm population, 3, 68, 72; growth of, 3, 140, 185; and independence, 115; jobs in, 110; lawn fertilizer and weed killers in, 136–37, 156; life in, 186; time spent with neighbors in, 64; upbringing in, 34, 37. *See also* farm towns; towns

city people, 129; knowledge of farming among, 6, 136; and neighbors, 54–55, 87, 101; view of farmers among, 110–11

clergy, 73–74, 86–91, 208n2, 209–10n6; and attendance at church, 88; as community leaders, 89–90; as connection with outside world, 90; difficulties in attracting and supporting, 82; foreign, 90; as missionaries, 90; and preaching and teaching, 83; sermons and lessons of, 90, 91; urban, 90. *See also* churches; religion and faith

collective bargaining, 166

college education, 17, 45, 99, 101, 183; acquisition of, 18, 23; as business asset, 31; as desired for children, 42; and economic constraints, 97; and experience with farmers, 51; and farm business, 104; and independence, 104, 105, 108, 109, 111, 113, 117; and interviewees, 195–96, 203n6; and land, 136; and religious values, 7; and women, 34, 35, 104

colleges, 68

commodities, 67, 168

commodities futures, 168, 172

common good, 95, 187, 188

communities, 16; civic betterment of, 64; and competition from outsiders, 57; and corporations, 178–79; declining population of, 3, 8, 52, 73; declining sense of,

46; ethos of, 117; family tradition in, 18, 43; and farm towns, 68–69; and formal organizations, 62–65, 66; and generational succession, 43–44; history and familiarity of, 47; immigrants in, 59–62; leaders in, 5; loyalty to, 54; narratives about, 47; neighbors in, 7, 9; religion in, 9–10; reputation in, 40; sense of, 52–53. *See also* churches; neighbors; social networks

competition: from corporations, 137, 138, 169, 182, 187; formal organizations as channeling, 65; for land, 55–58; with neighbors, 2, 7, 47, 51, 55–58, 110, 114, 117, 189, 206n6; from nonfarm investors, 58; with outsiders, 56–57; and personal relationships, 57–58; and specialization, 58; and technology, 141–42; from wealthy farmers, 56–57, 58

ConAgra, 177

construction work, 22

contractual relationships, 26–27, 29–31, 53. *See also* partnerships

corn: and ethanol, 158–59, 170–71; and GMOs, 152

corn growers, 6, 12, 13, 40, 50, 62, 193, 195

corporations, 24, 134, 169; agribusiness, 169, 188; agricultural, 187, 215n8; careers in, 188; competition from, 137, 138, 169, 182, 187; and co-ops, 62–63; effect of on land, 178; family, 5, 24, 27, 53, 109, 179, 182, 203n7; and family farms, 177–79, 187; and government policies, 137, 155, 182; investment by, 189; and markets, 163; and research, 5; and sense of community, 178–79; and traditional values, 177–78

cotton growers, 5, 6, 13, 19, 24, 65, 193, 194, 195; and GMOs, 152, 153; and immigrants, 61; and international markets, 165; local and regional associations of, 62; regional differences among, 166–67

cotton growers' associations, 166

crop insurance, 180, 181, 182

crop loss, 19, 22, 77, 180

dairy farmers, 5, 6, 13, 28, 63, 75, 154, 193, 195; and efficiencies of scale, 176–77; and immigrants, 60; and prices, 166. *See also* cattle business

daughters, 14, 17, 24, 25, 29, 42, 56, 108, 160, 188. *See also* partnerships; women

daughters-in-law, 34, 35, 36–37, 38

death: and choice of farming, 97; and churches, 83–84; and faith, 73, 76–77; and family conflicts, 32; and help from neighbors, 48, 49, 50; and meaning of farming, 20–21

debt, 23, 29, 111, 114, 171. *See also* banks; loans; money

Dirty Thirties, 21

economies, shift from agrarian to industrial, 3

education, 141; and careers, 42; and choice of farming, 97; of interviewees, 195–96. *See also* college education; schools

efficiencies of scale, 176–79, 187, 189

Emerson, Ralph Waldo, 78

employees/hired hands, 22, 24, 27, 35, 39, 53, 103, 147, 149, 161; immigrants as, 59, 60, 61; students as, 50, 51. *See also* labor

Environmental Working Group, 194

ethanol, 158–59, 170–71

ethnic diversity, 47–48, 59, 60, 61. *See also* immigrants

evangelicals, 80, 85, 89

faith. *See* religion and faith

families, 12–45, 192; aging of, 3; and business, 7, 8, 9, 13, 24–27; and business model, 29–30, 31, 32, 108–9; and choice of farming, 97; complexity of relationships in, 203–4n8; conflict resolution by, 29–33; conflicts in, 9, 27–33, 93, 108; conflicts in cautionary tales about, 31–32; conflicts involving women in, 35–38; and connection with previous generations, 9, 14–15, 16; and contractual relationships involving money, 30–31; criticism of farmers among, 110; and death, 32; as farming same place for generations, 1, 15, 16, 44; generational differences in, 29; and generational succession, 43–45; hardships of, 19–23; and independence, 108, 116, 117; and information technology, 150; and inheritance of farm, 33; intergenerational, 12, 44; and knowledge, 44; labor by members of, 12, 13, 24, 27, 43–44, 189; lack of extended members of, 53–54; and land, 1, 8, 25, 120, 121–26, 133, 135, 139, 179, 212n3; love for, 187; and machinery, 145–47; and machinery ownership arrangements, 30; machinery sharing by, 14, 17; and markets, 164; and nearby relatives, 32; patrilocal, families (*continued*)

36; relations of as integral to farming, 1; rivalries in, 27; stories of, 4, 5, 6, 8, 9, 11, 14–16, 19, 20, 21, 22, 23, 171, 192 (*see also* family traditions); and technology, 160; and third-party mediation, 33; and transition agreements, 30; traveling and visits with, 54; and trust-building, 29; as units of consumption and production, 13; valuing of, 12–13; as working together, 27. *See also* ancestors; children; fathers; husbands; parents; wives

family corporations, 5, 24, 27, 53, 109, 179, 182, 203n7

family farms, 3, 5; and business, 24, 31, 32; change and future of, 42; and corporate interests, 187; and corporations, 177–79; and family relationships, 12; and farm subsidies, 181, 187; incorporation of, 179; and machinery, 145–47; and market, 181; and neighbors, 55; and older farmers, 43, 44; and past, 14, 22; and regulation, 137; and size of farms, 187; and technology, 142, 157, 160, 161, 162

family therapists, 33

family traditions, 13–19, 21, 22, 44–45, 187; in communities, 18, 43; and flexibility, 169–70; and place, 15–16; and previous generations, 13; and risk, 172; and self-reliance, 106; and skills, 16; and spatial connections, 15, 16; and stories, 14–16

farm association meetings, 54

Farm Bureau, 62

farmers: diversity of, 192–93; generational continuity of, 14–15; population of, 3, 8, 10, 43, 46, 68, 72–73, 80, 87, 90; public knowledge about, 3, 6, 110–11, 185–88, 216n1, 216n2; stereotypes about, 3–4, 166, 170, 171. *See also* business; families

farmers, older, 43–45, 151; and faith, 76–77; and family business, 28, 30, 44; and independence, 117; and land, 123; as unwilling to let go, 30; valuable advice from, 28, 66; valuable interactions with, 70–71. *See also* aging; fathers; grandparents; mothers; parents

farmers, younger: and conflict with older farmers, 28, 30; and family networks, 25; as following in parents' footsteps, 25; income and investments of, 44; independence of,

117; parents' assistance to, 43–44. *See also* children; daughters; sons

farmers' co-ops, 15, 62–64, 65, 73, 84, 150, 155, 167

farming: and alternative careers, 22–23, 97–100, 105, 106, 116, 124, 188; American history as rooted in, 3; as in the blood, 37, 99, 135; changing nature of, 2, 28, 40–41, 42, 43, 47, 51–54, 55, 135, 187 (*see also* technology); choice of, 2, 17–19, 36, 42–43, 97–101; diversity of, 8; and industrialization, 3; large-scale industrialized, 121; multigenerational, 28; no-till, 53, 132, 133, 136, 147–48, 160, 162; passion for, 186; as practice, 115

farms: and cities, 140; and families in same place for generations, 1, 15, 16, 44; larger scale, 53, 81, 109, 117, 144, 177, 181–82, 192, 203n8; multigenerational, 5; size of, 6, 53, 187, 195, 203, 213n2

Farm Service Agency (FSA), 137, 147

farm subsidies, 179–84, 187

farm towns, 68–69; businesses in, 69; declining population of, 59, 62, 68, 69; immigrants in, 59; populations of, 207n13. *See also* towns

fathers: death of, 27, 32; land ownership by, 28–29; ownership of machinery by, 29; retired, 26; traditions passed down from, 17–18. *See also* farmers, older; parents; partnerships

Federal Deposit Insurance Corporation (FDIC), 20

fertilizer, 10, 140, 158, 162, 165, 187. *See also* chemicals

fiber security, 180, 181

food policy, 7

food prices, 4

food safety, 141, 155–56

food security, 180, 181

food stamp programs, 3, 181

formal organizations, 9, 48, 62–65

fundamentalists, 75, 93

funerals, 83–84, 90

futures markets, 161, 168

gender, 3, 9, 13, 24–25, 204n13. *See also* men; women

genetically engineered seed, 2, 10, 140, 141, 172, 213n2

genetically modified organisms (GMOs), 152–53, 154, 155, 157, 160

genetic engineering, 152–57, 167, 171. *See also* technology

globalism, 182–83. *See also* markets

global positioning systems (GPS), 147–49, 161, 172

gossip/rumors, 49, 66, 110, 111. *See also* neighbors

government: and crop insurance programs, 180; distrust of, 182, 183; energy policy of, 3; and environmental policy, 3; and farm subsidies, 4, 179–84, 187; and information about farming, 182; and markets, 165–66, 168, 175, 183; questions about, 5

government regulations, 4, 213n2; and independence, 155, 156; and information technology, 150

grandparents, 14, 16, 18, 27, 123, 169; assistance from, 25, 43, 44; connection with, 14; farm labor provided by, 43–44; hardships of, 19–20; and intergenerational continuity, 14; learning from, 20; memories of, 39; work of, 34, 39. *See also* aging; ancestors; farmers, older

Grange, 84

Great Depression, 19–20

growers' associations, 167, 173

Guatemalans, 60

hardships, 9, 19–23, 114, 123, 189. *See also* accidents; illness; injuries; money

hard work, 98, 144, 187; and children, 40; and faith, 85, 87; and farmers' public image, 186; and food stamp programs, 181; and immigrants, 61; and land, 125, 126; and neighbors, 50; and success, 110, 112, 114

harvest, 24–25; help with, 44, 48, 87; typical day of, 25; women's work during, 34, 35

health insurance, 35

herbicides, 152, 153, 160, 162, 213n2. *See also* chemicals; pesticides

heritage associations, 63

Hispanics, 60, 78, 85

husbands: absence of, 35, 42, 146; and conflicts with wives, 9; fieldwork by, 24–25; as interviewees, 195; and wives, 5. *See also* men; wives

husband-wife farms, 24–25, 203n7

husband-wife-plus arrangements, 203n7

identity, 16, 44, 98, 99–100

illness, 9, 19; and choice of farming, 97; and churches, 83; and faith, 73, 76, 92; and GMOs, 152; and help from neighbors, 48, 49, 51, 87. *See also* hardships; injuries

immigrants, 59–62, 93

immigration, 47–48, 60

implement dealers, 23, 69

independence, 8, 10, 75, 95–118; as being own boss, 10, 95, 96, 97, 100, 102–9, 113, 115, 116, 118, 138, 162; challenges of, 100, 107–8, 112–13; changing senses of, 117; and character, 117; and childhood development and socialization, 96; and choice of farming, 95, 97–101; and cities, 115; complexity and malleability of personal, 96; and day-to-day activities, 102–9; and decision-making, 102, 104–6, 111, 113, 162, 189; and diversity of tasks and skills, 102, 106–8, 211n6; and equipment, 117; and faith, 116; and family, 116; and family relationships, 108, 117; and farming as practice, 107; and farm subsidies, 181; and GMOs, 155; and good work, 103; and government regulations, 155, 156; and hardship, 114; and identity, 100; and individualism, 101; and introverts vs. extroverts, 96; and knowledge, 111; and limited social contact, 101, 102; and markets, 116, 117, 164, 170; and neighbors, 101, 116, 117; and niche marketing, 172; and obligations, 100, 101; and obligations to parents, 28–29, 108; perceptions of, 97; and responsibility, 103, 111–13, 115–18; and risk-taking, 112–13; and self-governing principles, 115; and self-interest, 96, 117; and self-realization, 113; and self-reliance, 106–7; and social landscape, 97; and social networks, 96; and social norms, 116; and success and failure, 10, 109–15; and technology, 97, 117, 161–62; threats to, 116–17; value of, 187; and weather, 116; and working alone, 101, 102–3, 108–9

individualism, 112; American, 10, 95; checks on, 116; expressive, 96; and independence, 101; and social relationships, 116

industrialization, 140

223

industries, 13, 185
information, sharing of with neighbors, 46,
47, 49, 65–68, 116
information management, 134
information technology, 2, 10, 140–41,
149–52, 162; and business activities, 150,
151–52
injuries, 19, 20, 49, 75. *See also* accidents;
hardships; illness
in-laws. *See* partnerships; women
interest rates, 20
international networks, 54
international trade, 3
Internet, 66, 67, 149, 150–51, 161
interviewees. *See* research
investors, 58, 126, 182
in vitro fertilization, 157

Jackson, Andrew, 3
Jefferson, Thomas, 3

Kiwanis, 63

labor, 13–14, 22, 120; business arrangements
for, 30; exchanges of for land and equipment use, 25; by family members, 12, 13,
24, 27, 43–44, 189; immigrant, 59, 60;
paid, 188; sharing of, 26, 53; and technology, 162. *See also* employees/hired hands;
hard work; off-farm jobs; towns, work in
labor market, 7
land, 8, 119–39; ability to acquire, 111–12;
aesthetic appeal of, 8, 126–27, 129, 130–31,
139; and ancestors, 122, 123, 128, 178;
attachment to, 120–21; and blood relatives, 37; and business, 121, 132; challenges
of, 130; changing relationships to, 47, 192;
and chemicals, 10, 136–37, 161; and competition with neighbors, 55–58; contamination of, 136–37; cost of, 59; and death
of parents, 32–33; in different locations,
179, 189; distance from, 7, 131–34, 136;
effect of corporations on, 178; exchanges
of labor for, 25; and families, 1, 8, 25,
120, 121–26, 133, 135, 139, 179, 212n3;
fathers' ownership of, 28–29; and feeling
of imprisonment, 124; and geographic
mobility, 212; and GMOs, 154; and God,
127, 138; and government regulation,
136–38; and hardship, 123; help for
children with, 55–56; and immigrants,

59; independence in ownership of, 95;
inherited, 14, 18, 26, 32, 36, 124, 125,
138; intimate relationship with, 121, 122,
129, 139, 154, 178, 189; and investors,
126; labor involved in working, 129; as living fragile thing, 136; long-term thinking
about, 135; and machinery, 10, 120, 121,
131, 133; and markets, 178; meaning of,
120; and memory, 123–24; mental relationship with, 129–30; monetary value
of, 138, 139; multigenerational use and
ownership of, 3, 16, 17–18, 19, 26, 120,
121, 122; and neighbors, 50, 125; newly
acquired, 135; ownership of, 108; and
partnerships, 26; price of, 47, 182; purchased, 124–25, 128, 135; recreational
uses of, 131; relationship with, 7, 10;
rented, 25, 27, 59, 116, 125, 135, 189;
and reputation, 125; sale of, 22; scarcity
of, 138, 188; sharing of, 26; and shoddy
business practices, 126; and solitude,
126–27; speculation in, 126; stewardship
of, 134–39; stories about, 15, 16, 122–23;
struggles to hold on to, 19; and sustainability, 121; and technology, 7, 120, 121;
tiling of, 128; unproductive, 124; and urban investors, 131; visceral vs. conceptual
understanding of, 132–34; working of as
meaningful, 126–30. *See also* farming
Latinos, 61, 88
Lions Club, 63
loans, 4, 18, 35, 167, 171, 176; from banks,
10, 21, 23, 25, 148; church help with, 84;
parents' help with, 25, 43
Lutherans, 75, 81, 83

machine and equipment dealers, 69, 150
machinery and equipment, 24–25, 143; and
ability to farm more land, 144; and aging,
44; ambivalence about, 160; and assistance
to neighbors, 50; and children, 41; conflicts over, 28; cost of, 2, 10, 47, 87, 116,
142, 165, 188; efficiency of, 142–43, 144,
145, 147–48; and faith, 88; and family
farming, 145–47; and family life, 145–47;
family ownership arrangements for, 30;
family sharing of, 14, 17, 25, 26; fathers'
ownership of, 29; fun of using, 143–44;
and immigrant labor, 59, 60; increased
work to pay for, 144–45; and independence, 117; information about, 67; labor

exchanged for, 25; and land, 7, 10, 120, 121, 131, 133, 192; mechanical skills and training for, 145; rental of, 30, 53; repair of, 187; risk from, 42; sharing of, 2, 14, 17, 25, 26, 47, 50, 53, 160; and technology, 140–41, 142–47; women's operation of, 35; work to pay for, 145

manufacturing sectors, 166

markets, 61, 163–84; and caring for families, 164; concern about, 10–11; as beyond control, 164–67, 168, 171, 176, 183–84, 187; decisions about, 164; effects of changes in, 3; and efficiencies of scale, 176–79; and farm subsidies, 181; and flexibility, 169–70; fluctuations in, 2, 10, 114, 116, 117, 165, 166, 167–68; and government policies, 165–66, 168, 175, 183; and independence, 164, 170; information about, 167; international, 8, 163, 165, 166, 167, 168, 171, 172, 180, 181, 183; and land, 178; and larger vs. smaller farms, 176–77; as manipulated, 163; niche, 172–76; and prices, 163, 165; as rigged against farmers, 165, 167–68; and risks, 171–72, 173; stories about, 21; strategies for dealing with, 169–72; and truck farmers, 50; unpredictability of, 165, 168, 171. *See also* supply and demand

marriage, 17, 24, 36–37

Masonic lodges, 84

materialism, 29, 49–50

meat processing facilities, 47, 59, 61

media, 4, 6, 110–11, 180, 186, 216n1

mediation, third-party, 33

men: absences of, 35, 70; and family traditions, 17–18; as interviewees, 5, 195; and socializing, 70; and Sunday school and Bible study groups, 83. *See also* fathers; gender; grandparents; husbands; parents

Mennonites, 75

mentors, 51, 100

methane, 159

Methodists, 75, 77, 81, 83

Mexicans, 60, 61

Mexico, 156

migrant workers, 60

miracles, 75, 77

money, 34–35; care about handling, 29; and choice of farming, 97; egg, 35; and familial relationships, 30–31; free, 114; and land, 178; pin, 35; problems with, 19–20, 21.

See also debt; hardships; loans; machinery and equipment, cost of

Monsanto, 161, 177

Mormons, 85

mothers: roles of, 33–38; traditions passed down from, 17–18. *See also* parents; partnerships; women

mothers-in-law, 35, 36, 37

Muir, John, 78

narcissism, 96

neighborliness, 47, 48, 52–53; and formal organizations, 9; and independence, 116

neighbors: blood relatives as, 47; and changing nature of farming, 51–54, 55; competition with, 2, 7, 47, 51, 55–58, 65, 110, 114, 117, 189, 206n6; contact with, 52–54, 61–62; and co-ops, 62–63; ethnic diversity of, 47–48; family histories of, 113; and formal organizations, 48, 64; helpfulness of and sharing with, 47, 48–51, 52–53, 55, 58, 87; immigrants as, 59–62; and indebtedness and reciprocity, 49; and independence, 101, 117; information sharing with, 46, 47, 49, 65–68, 116; lack of privacy among, 110, 111; and land, 50, 125; and legal and financial considerations, 53; and new machinery, 144; partnerships with, 26; and personal responsibility, 50; questions about, 5; relationships with, 9, 46–71; rural vs. urban, 50; and self-interest, 49–50; and self-sufficiency, 51; and sense of place, 47; socializing with, 70–71; and social norms, 47, 49–50; successes and failures of, 113–14; time spent with, 52–54, 64–65; urban vs. rural, 54–55; value of, 46; and work ethic, 50. *See also* churches; communities; gossip/rumors

newcomers, 61–62, 87

nuclear family, 24

nuclear-family-plus, 24

off-farm jobs, 23, 24, 34–35, 36, 114, 176, 203. *See also* labor

organic farming, 174–75, 215nn5, 6

parents, 12; absence of from children, 42; assistance from, 43; business consultations with, 25; children as working alongside, 38; and children's choice of farming, 17–18, 97, 100; children's conflicts

parents (*continued*)
with, 9; connection with, 14, 16; death of, 32–33; help from, 24, 25, 43, 44, 55–56; help from to start farming, 29, 160; and land, 122, 123, 124; land rented from, 25; learning from, 20; living close to, 188; and love of farming, 23; obligations to, 28–29, 108; respect for, 40; retired, 24, 25, 26; as role models and mentors, 100; wisdom of, 14; women's care for aging, 37, 38. *See also* ancestors; farmers, older; fathers; grandparents; mothers
partnerships: with blood relatives vs. in-laws, 26; brother-brother, 24, 25, 26, 27–28, 29, 37, 203n7; complexity of, 26; father-daughter, 24, 25, 203n7; father-son, 1–2, 24, 25–26, 28–29, 30, 187, 203n7; father-son-in-law, 18, 24, 25, 28; formal, 13, 24; informal, 13, 24; and inherited land, 26; and intergenerational relationships, 43; involving uncles, 25; mother-son, 203n7; with nonrelatives, 26; sibling, 5; sister-brother, 203n7. *See also* business; contractual relationships
patrimony, 36
Pentecostals, 78, 89, 93
pesticides, 2, 10, 136, 137, 140, 141, 153, 156, 213n2. *See also* chemicals; herbicides
place, sense of, 15, 16, 18, 20, 39, 124
prices, 10, 171, 213n2; and crop insurance programs, 180; and farm subsidies, 180–81; fluctuation in, 168; of food, 186; and markets, 163, 165; speculation on, 168
price support programs, 181
principal operators, 203–4n8
producers' cooperatives, 173
Protestantism, 93
Protestants, 75, 89
public health, 3

race, 59
rationality, 105–6
religion and faith, 9–10, 72–94; and agrarian vs. industrial economy, 3; and clergy, 86–91; and contact with God's creation, 78–80; and creation vs. creator, 79–80; and denominational diversity, 85; and dependence on God, 88; among different generations, 73; different traditions of, 85–86; as divine support, 74–78, 79, 93; and ethnic traditions, 93; expressiveness about, 89; and immigrants, 93; and independence, 116; and interpretation of farmers' lives, 208n2; and kind vs. wrathful God, 78–79; and lack of control, 72–73, 74, 75–76, 88; and land, 127; as meaningful, 73; misgivings about, 91–94; and modern farming methods, 92–93; and multigenerational loyalty, 93; and naturalistic spirituality, 78; and relationship to God, 93–94; shifting patterns in, 88; and superstition, 77, 92–93; and technology, 7; in urban vs. rural areas, 208n1; and work on Sunday, 86; and young people, 77, 81. *See also* churches; clergy
research: analysis in, 197; anonymity in, 196; interviewees for, 4–6, 194–96, 203n6; interview methods for, 196; qualitative, 6, 191–97; quantitative, 191
responsibility, 95; and farmers' public image, 186; and food stamp programs, 181; and independence, 103, 111–13, 115–18; long-term, 135; and neighbors, 50; for success and failure, 109–10, 111–12
retirement/retirees, 1–2, 24, 25, 26, 68, 160; and churches, 85, 86; and clergy, 86; and debt, 29; farming past normal age of, 44; refusal of, 28. *See also* aging
risks, 9, 21; and ancestors, 172; and children, 39–40; and faith, 72, 73; and family conflicts, 29; and family traditions, 172; from farm equipment, 42; financial, 23; and independence, 112–13; and markets, 171–72, 173; necessity of taking, 23; and technology, 142. *See also* accidents; injuries
role models, 100
Rotary International, 63
Round Up, 152, 153

schools, 3, 64–65, 68–69, 73, 81, 150. *See also* education
seed: and agribusinesses, 134, 161, 173; genetically engineered, 2, 10, 140, 141, 172, 213n2; grown by farmer vs. purchased, 161; price of, 154; varieties of, 136, 156; weed-resistant, 162
selective breeding, 157–58
sharecroppers, 19, 188
siblings, 5, 24, 25, 27, 32, 36, 39, 97, 98, 110. *See also* brothers, rivalries between; partnerships

social networks, 66, 85; geographically dispersed, 54; and independence, 96, 112; and neighbors and competition, 206n6; and niche marketing, 175; and risk-taking, 112; and travel, 54. *See also* churches; communities; families; neighbors
soil conservation, 137
soil science, 172
sons: and father's land and machinery ownership, 29; traditions passed down to, 17–18. *See also* parents; partnerships
soybean growers, 6, 12, 13, 152, 193, 195
soybean markets, 168
sports, 81, 88
stories. *See* families; family traditions
success vs. failure, 109–12, 114
supply and demand, 163, 165, 168, 172. *See also* markets
sustainable energy, 158–60

technology, 10, 135, 140–62, 167, 170, 187, 189, 213n2; adaptation to new, 7; ambivalence about, 142, 160–62; and competitiveness, 141–42; cost of, 142, 151; effects of changes in, 3; and environmental concerns, 141, 155, 156, 160; expertise in, 155–56; and faith, 73, 88; and family farming, 142, 161; and family relationships, 150, 160, 161; and independence, 97, 117, 161–62; information about, 67, 157, 160; knowledge of, 186; and labor, 162; and land, 7, 120, 121, 192; and machinery, 142–47; and niche marketing, 175; and religion, 7; risks and rewards of, 142; and traditional values, 161; utopian possibilities of, 161; and values, 8. *See also* chemicals; genetic engineering; machinery and equipment
towns, 43, 132, 188, 191; activities in, 41; and churches, 72, 80–81; churches in, 82, 85, 86, 87, 89, 93; co-ops in, 62–63; and hard work, 112; immigrants in, 59, 61; jobs in, 24; life in, 39, 63; retirement to, 123; and schools, 42, 64; social contacts in, 101; work in, 12, 17, 24, 34, 36, 37, 56, 67, 110, 112, 176 (*see also* labor; off-farm jobs). *See also* cities; farm towns
townspeople, 50, 79, 88, 131, 174
trade unions, 166
traditions, 3, 7. *See also* churches; family traditions

Transcendentalists, 78
travel, 54, 68, 88, 124
truck farmers, 6, 27, 50, 60, 66, 167, 171, 193, 195
Tyson Foods, 177

urban areas, 3, 43, 59, 101. *See also* cities; farm towns
US Department of Agriculture (USDA), 4, 27

vegetable growers, 13, 193

wageworkers, 103, 166. *See also* towns, work in
weather, 8, 22, 114, 172; and faith, 72–73, 75, 76, 78, 88; and hardships, 19, 20; and independence, 116; lack of control over, 128; and media, 4; stories about, 21; uncertainties about, 10, 35
wheat growers, 5, 6, 193, 194, 195; associations for, 62, 167; and harvest, 24–25; and markets, 168; and neighbors, 50
wind energy, 159–60
wives: and conflicts with husbands, 9; errand running and equipment operation by, 24–25; as interviewees, 195–96; and machinery, 146. *See also* husbands; women
women, 2, 37–38; and business arrangements vs. family activities, 36; and care for aging parents, 37, 38; careers of, 34; and childhood expectations of life, 36; with college degrees, 34, 35, 104; and cooking, 36; as distant from parents and siblings, 36; and family activities, 36; and family traditions, 17–18; and farmwork, 24–25, 34, 35; and formalized arrangements, 36; and husbands' absence, 35; income of, 34–35; independent identity of, 34; and in-laws, 34, 35, 36–37, 38; as interviewees, 5, 195; and off-farm jobs, 34–35, 36, 204n12; as outsiders, 36, 37; as raised on farms, 36; roles of, 33–38; and socializing, 70; sources of conflict for, 35–38; and special monies, 34–35; and Sunday school and Bible study groups, 83. *See also* daughters; gender; wives

young people: and changing farming practices, 28; and church attendance, 81, 88–89; and farming as career, 14, 187–88; independence of, 115. *See also* children; daughters; farmers, younger; partnerships; sons

CPSIA information can be obtained
at www.ICGtesting.com
Printed in the USA
JSHW031442120720
6635JS00005B/13